Burns-

Good to meet you, hope to work with you in the future

Hal

Harold Frediani, M 7/5/05

Power Plant Permitting

POWER PLANT PERMITTING

BY

HAROLD A. FREDIANI, JR.
P.E., P.H.

AND

KIMBERLY MASTERS EVANS
P.E.

PennWell Publishing Company
Tulsa, Oklahoma

PennWell Publishing Company
1421 South Sheridan/P.O. Box 1260
Tulsa, Oklahoma 74101

Library of Congress
Cataloging-in-Publication Data

Harold A. Frediani, Jr.
Power plant permitting / byHarold A. Frediani, Jr. and Kimberly Masters Evans
p. cm.
Includes index.
ISBN 0-87814-592-3
1. Electric power-plants--Location. 2. Electric power-plants--Environmental aspects. 3. Electric power-plants--Licenses--United States. I. Kimberly Masters Evans II. Title.
TK1005.E93 1996
333.79'32'0973--dc20 96-44184
CIP

Printed in the United States of America
1 2 3 4 5 00 99 98 97 96

Contents

Tables

FIGURES

Foreword

There is no other endeavor quite like permitting a power plant. It requires a blend of engineering, scientific, sociological, business and legal expertise to pull it all together. Even in its simplest form there is a maze of regulations to muddle through. Let us assure you that it is a political process. You may encounter intra-agency conflicts and jealousies. Agencies you have never even heard of will come out of the woodwork and demand to have a say. You may encounter vocal well-organized public opposition.

As consultants we have been in the middle of the fray—trying to balance a multitude of people with varying agendas. There is the client who wants to build an economically viable plant in the face of ever-tightening regulatory demands. The fickle public needs and demands more electricity but no one wants the plant in their backyard. Environmental groups almost always oppose power projects. We once attended a certification hearing for a 684 MW repowering project where a Greenpeace representative showed up and complained bitterly because the free coffee was being served in Styrofoam cups. You never know what you are going to run into.

We wrote this book because we have been carrying this information around in our heads for years and wanted to record it for posterity. Although specific regulations vary by location, the overall process of permitting a new or upgraded power plant has many universal features—picking a suitable site, scoping environmental problems and solutions, and designing a well-engineered project. We hope this book makes you better prepared to do these tasks or oversee the people who are performing them. We have included bits of advice and real-life stories wherever possible to put information into perspective. (Kim's note—Hal has been regaling his junior staff with these stories for years and it is good to finally see them in print. They are a testament to his immense expertise in the field, and of course, prove that he knew Thomas Edison personally.)

The introductory chapter gives an historical background of the power industry and the changes it has undergone over the years. Each of the other eight chapters addresses a particular aspect of the permitting process; they were written to be stand-alone chapters as much as possible since projects have different areas of concern. Taken as a whole the book will give you a very broad overview of the multiplicity of tasks involved in permitting a power plant project. Please write to us in care of Pennwell Publishing with any comments or suggestions for our next edition.

Sincerely,
Harold A. Frediani, Jr.
Kimberly Masters Evans

Dedications

I dedicate this book to my parents, Jackie and Morrison Masters, for their motivation; and to my sisters Robin, Jill, and Audrey for their encouragement; and to my husband Weldon Evans for his understanding.

Kim Masters Evans

I dedicate this book to my wife, Judy, and to my children Miriam, Lisette, and Harold III, who all went for months with no husband or father to make time for me to work on this book.

Harold A. Frediani, Jr.

Introduction

A power plant is unique in the world of industrial production. No other facility produces an intangible product, energy rather than matter. You can't put this product in a container and sell it in a store. You have to produce it as it is needed and deliver it instantaneously for it to be useful. You have to produce it 24 hours a day and seven days a week, in greatly varying amounts. With few exceptions (e.g., pumped storage or compressed air facilities), you have to be able to produce the maximum amount your customers could ever want at any given time that they might want it.

Because of this uniqueness, a power plant also has unique effects on the environment. Essentially, a power plant converts potential energy (stored in fuel, air or water) into electrical energy. The raw ingredients—most often fuel and air—are transformed into waste products, and the electrical energy is created in the process. These waste products are unused fuel and combustion residues, either gaseous or solid. Other waste products are secondary in nature, such as water treatment wastes and spilled fuels. The other waste usually produced is waste heat, either as heated air or heated water. Heated air is the result of combustion, and heated water is the result of condensing steam to make the generation process more efficient.

In most industries, raw materials are somehow processed into a finished product. Any waste produced is a result of inefficiency; therefore, waste minimization has its own reward in decreased cost of product. Power plants are similar in that a more efficient generation process consumes less fuel and so produces less fuel-derived waste products. Similarly, if more of the potential energy entering as raw material is converted to electricity, less waste heat is produced.

Secondary waste production operates differently. For example, a more efficient wet limestone SO_2 scrubber produces more waste sludge than a less efficient one. It is not just converting the waste from SO_2 to gypsum; it is also converting the limestone to gypsum. This whole process does nothing to enhance the electrical production; in fact, it makes it less efficient.

Because power plants and their impacts on the environment are unique, so is their permitting. The purpose of this book is to document the reasoning behind the regulations and to explain how such permitting can best be accomplished.

ACRONYMS AND ABBREVIATIONS

ac	Alternating current
AP- 42	An EPA document of air pollution emission factors
BACT	Best Available Control Technology
BOD	Biological Oxygen Demand
Btu	British thermal units
CAA	Clean Air Act
CAAA	Clean Air Act Amendments of 1990
CC	Combined cycle
CEMS	Continuous Emission Monitoring System
CEQ	Council on Environmental Quality
cfs	Cubic feet per second
cfm	Cubic feet per minute
CFR	Code of Federal Regulations
CO	Carbon monoxide
CO_2	Carbon dioxide
COE	U.S. Army Corps of Engineers
COS	Carbonyl oxysulfide
CT	Combustion turbine
CWA	Clean Water Act
dBA	Decibels (A-weighted level)
dc	Direct current
DDT	Dichloro-diphenyl-trichloroethane
DO	Dissolved oxygen
EA	Environmental Assessment
EID	Environmental Information Document
EIS	Environmental Impact Statement
EMC	Electric Membership Cooperative
EMS	Environmental Management System
ENU	Elementary Neutralization Unit
EPA	U.S. Environmental Protection Agency

ESA	Environmental Site Assessment
ESP	Electrostatic precipitators
FERC	Federal Energy Regulatory Commission
FEMA	Federal Emergency Management Agency
FGD	Flue gas desulfurization
FNSI	Finding of No Significant Impact
FWS	U.S. Fish and Wildlife Service
Genco	Generating company
GIS	Geographic Information System
HAPs	Hazardous air pollutants
HF	Hydrogen fluoride
HRSG	Heat recovery steam generator
IOU	Investor-owned utility
IPP	Independent power producer
ISO	International Organization for Standardization
kW	Kilowatts
LAER	Lowest Achievable Emission Rate
LNG	Liquefied natural gas
MACT	Maximum Available Control Technology
MOU	Memorandum of Understanding
MSL	Mean sea level
MSW	Municipal solid waste
MW	Megawatts
NAAQS	National Ambient Air Quality Standards
NEPA	National Environmental Policy Act
NESHAPS	National Emission Standards for Hazardous Air Pollutants
NGO	Non-governmental organization
NGVD	National Geodetic Vertical Datum
NOAA	National Oceanic and Atmospheric Administration
NOI	Notice of Intent
NO_x	Nitrogen oxides
NPDES	National Pollutant Discharge Elimination System
NRC	Nuclear Regulatory Commission

NSPS	New Source Performance Standards
NUG	Non-utility generator
NWS	National Weather Service
PCB	Polychlorinated biphenyl
PM	Particulate matter
POTW	Publicly Owned Treatment Works
PPP	Pollution Prevention Plan
PSC	Public Service Commission
PSD	Prevention of Significant Deterioration
PTE	Potential to emit
RACT	Reasonably Available Control Technology
RBW	Receiving Body of Water
RCRA	Resource Conservation and Recovery Act
REA	Rural Electrification Administration
RHA	Rivers and Harbors Act
RO	Reverse osmosis
SCR	Selective catalytic reduction
SCS	U.S. Soil Conservation Service, now known as the National Resource Conservation Service
SDWA	Safe Drinking Water Act
SIC	Standard Industrial Code
SIP	State Implementation Plan
SNCR	Selective noncatalytic reduction
SO_x	Sulfur oxides
SPCC	Spill Prevention, Control, and Countermeasures
SPDES	State Pollutant Discharge Elimination System
TBEL	Technology Based Effluent Limit
TDS	Total dissolved solids
TETF	Totally Enclosed Treatment Facility
TRC	Total residual chlorine
TSCA	Toxic Substances Control Act
TSS	Total suspended solids
TVA	Tennessee Valley Authority

UIC	Underground injection control
USFWS	U.S. Fish and Wildlife Service
USGS	U.S. Geological Survey
VOC	Volatile organic compound
WQBEL	Water Quality Based Effluent Limit

Chapter 1
Historical Background

Invention of the Central Generating Station

Thomas Edison is generally credited with the invention of the central electric generating station because he designed and oversaw the construction of the steam-driven Pearl Street Station in New York City in 1882. That plant transmitted direct current (dc) to mostly commercial users in the vicinity of the station. It supplied current sufficient to light 7,200 of the incandescent lamps Edison had previously invented in 1879. The first hydroelectric central generating station was placed in operation a short time later in Appleton, Wisconsin. That station, of about 12,500 watts capacity, produced enough current to power only 250 lights.

Edison believed that dc transmission was the best means of delivering electricity. He felt alternating current (ac) was too dangerous and referenced its use in the electric chair as proof. He formed a company that eventually grew to become the General Electric Company and concentrated his efforts on dc power. It was George Westinghouse, a contemporary of Edison's, who coordinated the development of ac generation and transmission of power after forming the Westinghouse Electric Company.

The first ac central generating plant was installed in Oregon City, Oregon, in 1890. It had a capacity of 480 kilowatts (kW), but more importantly, transmitted the power over a 14-mile transmission line to Portland, Oregon.

The technological conflict between Edison and Westinghouse over dc versus ac power was known as the "War of the Currents." Although Edison continued to insist on the superiority of dc power, his board of directors overruled him, and by 1896 a cross-licensing agreement between the two companies had been reached allowing each to use the patents of the other. Both went on to develop ac systems.

Development of Investor-Owned Utilities

After the turn of the century, a handful of large companies grew to dominate the electricity business. Each one had its own generating stations, transmission lines, and distribution facilities. In order to avoid the waste that would occur if competitive stations built multiple power lines to the same group of customers, local and state governments legislated the franchising of electric utilities. As compensation for having their territories restricted, the utilities were guaranteed a "reasonable" profit and no competition within their territories.

Under these conditions, private (or investor-owned) utilities (IOUs) prospered and quickly grew to service all of the metropolitan areas of the country. A few governmental units recognized the potential for profit and took over the lucrative power supply business themselves. These were mostly municipalities, although the state of Nebraska is a notable exception. Nebraska is served entirely by publicly-owned-and-operated power facilities. However, most electricity was provided by investor-owned utilities up until the time of the Great Depression.

Development of Government-Owned Utilities

The financial hardships of the Great Depression brought a lapse in investments in infrastructure by private companies. The federal government stepped into the void to supply both the needed infrastructure and new jobs by initiating large construction projects. It was at this time that the Tennessee Valley Authority (TVA) was formed. TVA built a system of dams along the Tennessee River with a threefold purpose: to stop devastating flooding in the region, to improve navigational maneuverability, and to supply electricity via hydroelectric generating stations.

During World War II, the availability of this electrical power helped spur the government to build large-scale war research and weapons development facilities in the area. Soon the full hydroelectric capacity of the river was no longer sufficient to provide the needed power and TVA began construction of steam electric generating units. By 1960, fully 75 percent of TVA's power production was from steam electric units.

The same time period saw the birth of other government investments in the power industry including Hoover Dam, Grand Coulee Dam, and Shasta Dam. Also numerous multi-purpose projects including electric generating stations were constructed by the Army Corps of Engineers (COE), the Bureau of Reclamation, and the Bonneville Power Administration. Meanwhile the federal government instituted the Rural Electrification Administration (REA), which spawned development of Electric Membership Cooperatives or EMCs around the country to service rural areas the IOUs had found unprofitable.

Monopolies

From its birth in 1882 up until the early 1970s, the electric utility industry experienced phenomenal growth. Private investor-owned companies enjoyed the luxuries of high demand, no competition within their territories and sparse governmental regulations on how they operated their businesses. They developed a virtual monopoly on the electricity market. The rate of return to stockholders in these companies was set by law and was typically about 14 percent.

The average electric customer, however, had only one choice for a power provider and could not shop around for a better rate. A few industrial users had the capital and engineering acumen to design and construct their own power plants and did so when the investment was deemed worthwhile. Likewise, companies with political clout could obtain permission to run transmission lines to their facilities from a neighboring utility whose rate structure was more palatable.

Rate hikes were imposed by the utility companies to pay for the construction of new generating capacity when a need could be shown to the appropriate governmental agency. This was typically the respective state's Public Service Commission (PSC). Once PSC approval was obtained and the facility was on-line, rates were raised to the customer to recoup the investment and, of course, maintain the profit guaranteed by law.

THE ENVIRONMENTAL MOVEMENT

The turbulent decade of the 1960s spawned the birth of the environmental movement in this country which had long-reaching effects on the power industry. The movement can be traced in part to a book published in 1962 by Rachel Louise Carson. The book, entitled "Silent Spring," foretold the extinction of songbirds as an inevitable result of the indiscriminate use of the pesticide DDT. This book was a significant part of the cause of the modern environmental movement. Concern about the environment gained great strength on college campuses and soon spread across America. National organizations such as the Sierra Club, the Audobon Society, the National Wildlife Federation, and the Wilderness Society were formed and they lobbied Congress for new environmental laws.

In 1969, the National Environmental Policy Act or NEPA was passed. Essentially, this act gave the government the power to determine the acceptability of certain industrial actions which could affect the environment. The act was applicable when a company took an action involving federal government jurisdiction in some way. The government could then examine the possible environmental consequences of that action, compare them with other possible actions (including no action), and determine whether the action was acceptable from an environmental viewpoint. Applicants for permits had to provide information in sufficient detail to meet the requirements of the document used for analysis, which was called the Environmental Impact Statement (EIS) and was prepared by the agency. The document prepared by the applicant was called an Environmental Information Document (EID). If the information was insufficient, the applicant would not receive approval for the action.

NEPA also established a Council on Environmental Quality (CEQ) to coordinate environmental policy. In 1970, the year of the first Earth Day celebration, the Environmental Protection Agency (EPA) was formed from smaller agencies such as the Federal Water Pollution Control Administration. A section of the Code of Federal Regulations (CFR) was set aside for the EPA and CEQ to publish rules to enforce the environmental acts or statutes approved by Congress. This section of rules is commonly called 40 CFR.

Also in 1970, the Clean Air Act (CAA) was passed. This act established air quality standards for six major air pollutants: particulate matter (PM), sulfur oxides (SO_x), carbon monoxide (CO), nitrogen oxides (NO_x), lead, and ozone. The CAA also required the states to develop implementation plans (SIPs - see 40 CFR 52) to implement and maintain these standards, which are known as National Ambient Air Quality Standards (NAAQS). All of these chemicals are emitted by fossil-fueled boilers which are covered under 40 CFR 60. The CAA requires a power plant to have a permit for each of its air pollution sources in order to discharge pollutants into the air.

In 1972, the Clean Water Act (CWA) was passed. The greatest impact of this act to the power industry was implementation of the National Pollutant Discharge Elimination System (NPDES) permit program. This program requires industrial facilities to obtain a permit for all piped discharges to surface waters, and to monitor those discharges. Rules for this program for power plants were encoded in 40 CFR, in particular in Section 423, which was written for steam electric power plants. Section 423 identifies typical power plant wastewater streams and sets limits on what EPA perceives to be the worst pollutants included in those streams. These end-of-pipe limits were based on the technology available to remove the pollutant in question, and are known as technology-based effluent limits (TBELs).

The CWA also required states to define water quality classifications and standards, based on the intended use of each water body in the state. These water quality standards are used to set water quality based effluent limits (WQBELs). The NPDES program was structured so that individual states could take over its administration, and most states have done so.

In 1974, the Safe Drinking Water Act (SDWA) was passed. As a result, EPA set Primary and Secondary Drinking Water Standards, limiting the levels of contaminants in water that could be provided to the public as drinking water. Most states used these standards to set acceptable levels in groundwater. These limits serve to restrict concentrations of pollutants in power plant discharges to groundwater in both NPDES permits, and in solid waste disposal permits.

In 1976, Congress passed the Toxic Substances Control Act (TSCA) and the Resource Conservation and Recovery Act (RCRA) to deal with hazardous wastes issues. Results of these two acts include a ban of PCBs in any NPDES discharge and requirements for the handling and storage of hazardous wastes. For power plants this legislation is applicable for acids, caustics, solvents, and waste paints.

The electric industry was affected greatly by the legislation arising from the "environmental movement." The permitting process became a maze of laws and regulations which touched almost every aspect of operation from fuel choice to heat dissipation. In order to meet the ever-changing regulations, the industry has relied upon a host of pollution control devices, new engineering designs and, frequently, environmental consultants; all of which are expensive options.

DEREGULATION AND COMPETITION

As the political mood changed from liberal to conservative during the 1980s, a trend toward deregulation of industry emerged and this trend is still popular today. As went the telephone and airline industries, so the electric utility industry is going. Utilities now find themselves competing with non-utility generators (NUGs) for the right to provide new generating capacity, and utilities' rate of return will soon no longer be guaranteed.

On April 24, 1996, the Federal Energy Regulatory Commission (FERC) finalized its rule requiring open access for all wholesale generators to transmit power on transmission facilities. Under this rule, FERC Order 888, a utility cannot charge an unassociated generator any more for power transmission than it charges itself. This kind of rule is virtually impossible to follow, or at least to prove that you have followed it. The inevitable end result of this rule is that utilities will be split, in an economic sense, into separate operating units, and eventually into physically separate companies, to provide generation and transmission/distribution. The net result for generating units will be to have them all owned by generating companies (Gencos) that will, for all practical purposes be NUGs.

The NUGs have taken a different approach to environmental compliance than that of traditional electric utility companies. In the past, utilities have fought environmental regulation on the basis that it infringes on their ability to produce cheap electricity and thereby obtain their guaranteed profit. Ultimately, the cost of that fight was passed on to the consumer. The NUGs on the other hand, have perceived environmental restrictions as just another cost of doing business. This has allowed them to license and build plants more quickly, thereby minimizing interest during construction.

By the early 1990s, the NUGs were agreeing to apparently costly environmental controls that had been fought by the traditional utilities. These controls included such then radical concepts as selective catalytic reduction (SCR) for NO_x control, and zero discharge systems for handling industrial wastewaters and stormwater. The new Gencos will also have to abandon their traditional confrontational role with regulators, if they expect to be competitive with the natural born NUGs. On the bright side, these Gencos will be considered as ordinary capitalist ventures rather than as infinitely deep pockets as they are presently conceived.

The remaining true monopolistic portions of utilities will then be just the transmission and distribution companies, the ones that will build, operate, and maintain transmission lines, switchyards, and other distribution facilities. The open access feature will require that these systems add additional lines and other facilities to accommodate added generating units wherever the Gencos decide to build them. The environmental permitting of these new transmission facilities will be a very difficult and tedious process, as it is today.

Deregulation and competition are results of the resurgence of the free enterprise system since the demise of Soviet Communism. This resurgence is spearheaded by the Republican Party, which was swept into control of Congress in 1994. These same Republicans made a serious attempt to dismantle the environmental regulations which they perceived as an impediment to free enterprise, in much the same way they perceived the regulation of utilities. However, the environmental movement had become so entrenched as to be unassailable, and the rolling back of environmental regulations could not be done.

The net result of all this will be to require NUGs/Gencos to try and locate new units at sites with existing units to take advantage of existing transmission capacity, or to site them at large industrial facilities which they are intended to serve. The former course allows for pollution offsets by retiring or cleaning up older units. The latter course has potential difficulty because the environment at these facilities is most likely already stressed. Again, the Genco is most likely to succeed at an industrial site if it can convince the industry to retire some of its older, less environmentally-friendly equipment. In short, the siting of new capacity under the new era of competition is likely to be very similar to that of the NUGs during the early 1990s, and will require that each project show a net environmental benefit in order to be licensable.

Today, all utilities must minimize cost wherever possible to remain competitive. Environmental compliance is a complicated endeavor and mistakes can be very costly. Likewise the cost of fighting limitations can be greater than that of complying with them. The purpose of this book is to provide insight and advice about how best to navigate the power plant permitting process now and on into the twenty-first century.

Chapter 2
Site Selection Studies

Regulatory Background

Chapter 1 described the passage of NEPA in 1969 and the creation of 40 CFR to be the rules of the EPA and the CEQ. The EPA rules are in Parts 1 through 799, also known as Chapter I. The CEQ rules are in Chapter V, Parts 1500 through 1599. The CEQ rules are only written from Parts 1500 through 1517 and are much shorter than the EPA rules. Parts 1501 (NEPA and agency planning) and 1502 (Environmental Impact Statement) are the most relevant to power plants. Part 1500 states the purpose, policy, and mandate of the CEQ, which is essentially to see that all federal agencies properly evaluate the environmental consequences of their decisions. Part 1501 describes how the federal agency determines whether or not an Environmental Impact Statement (EIS) is necessary. It includes the case where "actions are planned by private applicants or other non-federal entities" and the later involvement of the federal agency is expected. 1501.2 (c) requires the agency to "study, develop, and describe appropriate alternatives to recommended courses of action in any proposal which involves unresolved conflicts concerning alternative uses of available resources." Part 1502 describes the EIS which may be required.

The main federal agency which normally gets involved with the environmental permitting of power plants is the EPA. The relevant EPA rules are in 40 CFR 6, Procedures for implementing the requirements of the Council on Environmental Quality on the National Environmental Policy Act. Part 6 lists the following types of projects for which an environmental review is required:

- wastewater treatment plant construction under the Grants Program,
- construction of a new source under the NPDES Program,
- research and development projects carried out by EPA's Office of Research and Development,
- solid waste demonstration projects undertaken by EPA's Office of Solid Waste and Emergency Response, and
- EPA facility support activities.

The only one of these activities under which a power plant usually falls is that of a new source under the NPDES Program. If dredge and fill activities are included, the United States Army Corps of Engineers (COE) becomes involved. If the plant owner is an EMC, the REA can be involved. The Department of Energy has been involved with demonstration projects, such as under the Clean Coal Program. The Nuclear Regulatory Commission (NRC) is involved for nuclear power plants. Each of these agencies has its own rules which are written to comply with the CEQ rules described above, so their rules generally include all of the same requirements as the EPA includes in 40 CFR 6.

40 CFR 6, Subpart A describes how EPA decides whether a proposed project will require a full-blown EIS, can be handled with a simple Finding of No Significant Impact (FNSI), or can be exempted from the environmental review requirements. 40 CFR 6, Subpart B describes what the EIS must include, should one be required.

If a proposed power plant action involves a new source for NPDES purposes (40 CFR Subchapter N, the industrial wastewater portion of the NPDES program), the regulator will perform an analysis to determine whether an EIS is required. If EPA has NPDES authority, the regulator will be an EPA person. If the state has NPDES authority, the regulator will be a state official. It is important to remember that the state was only granted authority because the EPA determined that the state rules included requirements at least as stringent as those of the EPA. Similarly, if the proposed project involves another federal agency, its rules have also been reviewed by EPA and found to be acceptable. Therefore, no matter who the regulator actually is, the intent is to meet the same requirements as do the EPA rules.

With the above in mind, it is obvious that the easiest way to permit a project is to structure it so that an EIS is not required. But to do this, it is still necessary to look at all the subjects of an EIS, including the "appropriate alternatives" mentioned above. In fact, the applicant is required to submit some sort of "environmental information document" (or EID, defined in 40 CFR 6.101 (d)) which must contain all the information the regulator needs to evaluate the proposed project. The alternatives to be examined by the applicant are listed in 40 CFR 6.203 (b):

- size of facilities,
- location of facilities,
- land requirements,
- operation and maintenance requirements,
- auxiliary structures such as pipelines or transmission lines, and
- construction schedules.

Based on this short list, EPA typically has demanded that applicants address the following list of topics:

1. alternative means of satisfying the need for the project,
2. alternative site analyses,
3. alternative processes and facilities, and
4. the no-action alternative.

Item 3 will be addressed in Chapter 4, Site Specific Design Alternatives. The remainder of this chapter addresses items 1, 2 and 4.

Plant Conceptual Design

The first step in performing a site selection study is to identify the size and type of unit(s) needed. A traditional IOU, EMC, or government-owned utility will normally look at its existing system, evaluate expected need for power in the future, examine the economics of various fuels and technologies, and decide on desired schedule, size, and generation technology for the new unit(s). The same process can be used to evaluate repowering of an existing unit within the system. Under repowering, an old boiler is replaced by a new one that connects to the existing steam turbine-generator and associated facilities.

An Independent Power Producer (IPP) planning a cogeneration unit will usually look first at what sites are available to select a viable host (another industrial facility to which the new unit can supply something other than electricity). Typically, steam is provided for the industrial user; however, innovative cogenerators have supplied heat from exhaust gases, and even CO_2 for a CO_2 producing plant. In this case, the unit size and type are usually dependent on the requirements of the host facility.

In any case, the site selection process is based on a perception of what kind of units will be added and a general geographic area where they will be located. The plant design will have to be complete in sufficient detail that the environmental effects of the plant construction and operation can be estimated sufficiently to determine whether potential sites are viable or not. The design is also used concurrently to determine whether a site is viable from both an engineering feasibility and economic standpoint. This process is what is known as a Fatal Flaw Analysis. Details of plant design required include:

- primary and backup fuel(s),
- size and type of generating units, and
- preferred heat dissipation/condenser cooling system.

Fuel

The vast majority of power generation in this country comes from the combustion of fossil fuels, either in a steam electric or combustion turbine configuration. The second most popular fuel is nuclear, with steam electric power generation. If the generation method selected is other than nuclear or fossil fueled, it will most likely be wind or solar power, hydroelectric, fuel cell, or steam electric using municipal solid waste (MSW). The fossil fuels include coal, oil, natural gas, and more recently, Orimulsion (an emulsification of bitumen in water, produced in Venezuela).

The selection of the fuel and generating technology will affect the site selection in several ways. The most obvious is the transportation of the fuel to the site. For example, if natural gas is the primary fuel, proximity to an existing pipeline is advantageous to a site for both economic reasons (it's expensive to build pipelines) and environmental reasons (pipeline construction has environmental impacts and requires additional permits).

The fuel is also directly related to the quality of the air emissions. For example, the combustion of coal or oil produces significantly more air pollutants for a given power production than would the combustion of natural gas. The fuel is also directly related to the production of combustion byproducts, be they useful byproducts or solid waste. The most obvious example is coal ash. The byproducts require handling and disposal whether they are a waste or a usable byproduct. Except for natural gas, the fuel is usually stored on-site in bulk quantities. This allows the utility to negotiate the price of fuel with some flexibility as to delivery date, as well as to allow deliveries during a normal 40-hour week even though the power generation

FIGURE 2-1 Coal Pile

FIGURE 2-2 Oil Storage Tank

is an around-the-clock, seven-days-a-week proposition. The coal pile (see Figure 2-1) or oil tank farm (see Figure 2-2) can comprise significant portions of the site, and greatly complicate the permitting of stormwater from affected areas.

SIZE AND TYPE OF GENERATING UNITS

The size of the generating units directly affects the required size of the site. In general, the bigger the units are, the bigger the associated systems (e.g., heat dissipation system, solid waste disposal system, fuel handling system) will have to be. The type of generating unit relates directly to several of the site characteristics that determine whether a site is feasible and, eventually, preferred. The predominant types of generating technologies that have been used on a commercial scale include steam electric (see Figure 2-3 for open air units and Figure 2-4 for enclosed units), combustion turbine (both simple [see Figure 2-5] and combined cycle [see Figure 2-6]), and hydroelectric (see Figure 2-7).

FIGURE 2-3 Open Air Steam Electric Generating Units

FIGURE 2-4 Enclosed Steam Electric Generating Units

FIGURE 2-5 Simple Cycle Combustion Turbine Generating Unit

FIGURE 2-6 Combined Cycle Combustion Turbine Generating Unit

FIGURE 2-7 Hydroelectric Generating Units

The steam electric power plants are the only ones that were considered when the NPDES effluent limits (40 CFR 423) were written. Steam electric plants can be broken down by the type of fuel used. The nuclear steam electric plants are no longer popular in the United States, despite the fact that they have essentially no emissions of air pollutants. Oil-fired steam electric plants can be built capable of burning natural gas and vice versa. Coal-fired plants can be built so only minor modifications are required to allow them to burn oil or gas. Present research is looking at ways to produce a gas from coal, which would make it easier to interchange these fossil fuels. Typically, gas-fired plants have fewer air emissions than oil-fired, which have fewer than coal-fired. Expressed another way, gas-fired plants require the least amount of air quality control equipment of all the fossil fuels, followed by oil and then by coal.

Combustion turbine units (CTs) can be in either the simple cycle or combined cycle (CC) configuration. In simple cycle configuration, CTs do not generate steam; therefore, they require no heat dissipation system for condenser cooling water. In CC configuration, there are steam turbine-generators and associated equipment and a heat dissipation system is required. CC units are in great favor nowadays because of two reasons. The first is that the initial capital cost to construct a CC unit is only about half that of a conventional steam electric unit. The second is that the CC configuration is much more fuel-efficient than a conventional steam electric unit. CC efficiencies approach 60 percent while fossil fueled and nuclear conventional units are only about 35 percent efficient. In other words, 60 percent of the energy in the fuel is converted to electricity in the CC units; therefore, they burn slightly more than half the fuel to make the same amount of power. In order to hedge against future cost increases for natural gas and oil, some IOUs have required that a site be large enough to support future coal gasification facilities and the associated coal handling and storage equipment. A site with this much room is increasingly rare.

Since they have no air emissions, nuclear-fueled units reject more waste heat than comparable sized fossil-fired steam electric plants. This is because no waste heat is leaving up the stack. Typical design for a nuclear plant is 7 million Btu/hour per megawatt. The comparable number for fossil-fired units is about 4.5 million, and for CC units, about 2.5 million. For simple cycle CTs, the number is zero.

Steam electric units also have a requirement for ultrapure water, as makeup to the boiler or heat recovery steam generator (HRSG). The latter device is what the boiler is called when it is used to generate steam for the steam portion of a CC unit. The amount of this ultrapure water required is generally in direct proportion to the heat rejection rate of the unit; thus the nuclear plant requires the most, and the simple cycle CTs require none. The traditional method to make this ultrapure water includes the use of demineralizers, which produce a wastewater, and sometimes water treatment methods that produce a sludge (solid waste).

The coal-fired units typically have the largest land requirements for a given unit size. There are several reasons for this. First, the quantity of coal required to be handled usually precludes storing it indoors; therefore, an outdoor coal pile is required. Second, the byproduct production from the coal is greatest because the coal has the most impurities. Traditionally, enough storage area on-site has been provided to landfill this ash.

Preferred Heat Dissipation/Condenser Cooling System

Steam electric generating units have a condenser to allow the recycling of boiler water. There are two main reasons for this. The first is that the condenser allows the steam turbine to run at greater efficiency by lowering the back pressure on it. This is accomplished by condensing the steam at a temperature lower than the 212 degrees Fahrenheit it would condense at if released to the atmosphere. The second reason is that it would be too expensive to make all that ultrapure water and then use it only once.

At the time the steam electric guidelines (40 CFR 423) were written, the predominant type of heat dissipation system was once-through cooling. In such a system, water was pumped from the water source (typically a river, lake, estuary, ocean, or sea) through the inside of the tubes of the condenser to condense the steam which was passed outside the tubes. After passing through the condenser once, the water was discharged, usually to the same water source from which it came. In general, all other wastewaters were mixed with this so-called "circulating water" and thus treated by dilution. However, the guidelines indicated that the "best" technology for heat dissipation was a "closed cycle" or recirculating cooling system—one in which a heat dissipation device has been added. This device was generally either a cooling tower of some type, or a cooling pond. The water would be cooled in the device and then recirculated back to the condenser. Most of these devices used evaporation of a small portion of the water to achieve the cooling. This evaporation, and other water losses, required that such systems would not be 100 percent "closed," but would require some makeup water to be provided. The rules did allow new units to be built as once-through units, if a successful demonstration could be made that such a system would not result in appreciable harm to the environment. Because such demonstrations are allowed under Section 316(a) of the Clean Water Act, they are known as 316(a) demonstrations.

The net result of all this is that a power plant typically requires moderate amounts of water for boiler makeup and either large amounts of water for cooling tower (or pond) makeup, or very large amounts of water for once-through cooling. Thus, most site selection studies include as a criterion proximity to a significant water body. Exceptions are typically only made where some other mitigating factor is present. An example is a minemouth coal-fired plant where there is no cost for fuel transportation; in such a case water could be piped to the site. There are instances in which air-cooled condensers (see Figure 2-8) have even been used, thereby eliminating the

FIGURE 2-8 Air-Cooled Condenser

need for cooling system makeup. These cases are rare.

Evaporative closed cycle heat dissipation systems also typically have had discharges to water bodies. Because the makeup water has chemical constituents dissolved in it, and evaporation leaves these solids behind, there is a tendency for them to build up. If their level builds up too far, they can precipitate out (commonly known as scaling) within the condenser tubes. To prevent this, some of the circulating water is "blown down," i. e. discharged from the system to take some of the dissolved solids out. This discharge is typically made to the same water body from which the makeup is taken. Thus, it becomes a chemical discharge as well as a thermal (heated) discharge. This is another reason that proximity to a significant water body is required.

Transmission Characteristics

There is no purpose served by generating electricity until it is transmitted to a user. Thus, the location of the plant relative to the nearest compatible transmission line is very important. Most utility transmission systems require careful balancing of where large blocks of power are input. If too much power is input in a small geographic area, the system can become unbalanced and lose stability. This is a problem in that it increases losses (which means more fuel has to be burned to get the same amount of power to the customer, resulting in more pollution and more cost to the utility) and also decreases reliability (a failure in a component means longer down times for more customers). Therefore, the utility must weigh the economic and environmental costs associated with the location of new units relative to their transmission system.

Transmission lines are very expensive to build and to maintain. Obviously, the closer a new unit is to an existing compatible transmission line, the better choice it will be. This is true both economically and environmentally, as the permitting of transmission lines can be very difficult. The primary impact for which transmission lines are generally faulted is that their construction often cannot avoid impacting wetlands. Wetland impacts trigger COE involvement under the Dredge and Fill rules (Chapter II of 33 CFR), and if the impact is significant, an EIS can be required under COE rules. EPA may also become involved as a reviewer since the two agencies have an agreement (MOU or Memorandum of Understanding) allowing such involvement.

FUEL DELIVERY

Fuel is the largest operating cost of any fossil fuel-fired power plant. The cost of delivery is generally the largest portion of that fuel cost. That is why the analysis of the potential fuel delivery system is an important part of any site selection study. Prudency requires that fuel be available through at least two delivery modes. Coal can freeze inside rail cars, oil tankers can run aground and sink, and miners or transport workers can go on strike.

Natural gas is typically only transported by pipeline. Oil can be transported by pipeline, by truck, by rail (occasionally, see Figure 2-9), and

FIGURE 2-9 Oil Train Unloading Facility

by waterborne carriers (barges or tankers). Coal is transported mainly by rail or by waterborne carrier. Orimulsion is transported the same ways as oil, and in fact, further discussions of oil herein can be assumed true of Orimulsion as well. MSW is transported the same ways as coal. Utilities prefer plants that can burn more than one fuel for reasons of reliability. IPPs generally borrow the capital to build their units, and the associated lenders who provide the financing also prefer plants that can burn more than one fuel. Both of these attitudes developed as a result of the Arab oil embargo in the 70s.

If the fuel is coal, the preferred delivery methods include rail and water-borne vessels, although for short distances conveyors can be economically competitive. Also, in locations where rail or water-borne service is unavailable, trucking of coal has been justified. Thus a site is most advantageous if it is close to existing rail lines, port facilities, and heavy duty highways.

Oil is generally delivered by water-borne vessel and/or by pipeline, although it too has been trucked. In the past, IOUs or government-owned utilities had sufficient capital to build their own fuel transportation facilities, such as fuel oil terminals with associated short pipelines to inland power plants. Nowadays, it is both more economical and more environmentally friendly to use existing facilities.

If the fuel is natural gas, the only proven delivery system is by pipeline. Power plants typically do not store natural gas on site; therefore, they usually have a backup fuel (normally distillate oil). In general, it is advantageous to let the fuel supplier permit, build, and operate the gas pipeline up to the site boundary. The power generator then makes use of the gas supplier's storage capabilities and provides redundant delivery by the use of a backup fuel. The backup fuel is usually distillate fuel oil, as most gas-fired generating units can also burn distillate with little or no modification.

If the site is to be a multi-unit generating facility, it may prove desirable to build future units to run on different fuels. If the site is to be a single-unit or one-time construction development, it still might be desirable to switch fuels later, either for economic or environmental purposes. An example would be the addition of natural gas to allow co-firing for NO_x reduction. This means that a site has an advantage if it is more amenable to all modes of fuel transportation.

SITE COMPARISON AND PREFERRED SITE

The intent of NEPA was to incorporate the analysis of environmental factors in with the engineering and economic factors. During the early years of the environmental movement, the typical utility performed its site selection process according to the old methods, evaluating engineering feasibility, reliability, and economics to determine a site, and then performing an environmental analysis to justify the choice. Now, it has been realized that it is a more cost-effective method to factor the environmental analysis in from the beginning. In most cases, a site with an environmental drawback can be made permittable by performing some sort of mitigation along with the project. For example, if a project requires the destruction of some wetlands, the proper mitigation might be to create new wetlands somewhere else (see Figure 2-10). By adding in the cost of the

FIGURE 2-10 Artificially Created Wetlands

required mitigation, the applicant can put the environmental analysis on an economic basis to allow comparison between sites.

Candidate Area Screening

There are numerous methodologies for starting with a region and working down to candidate site areas, then to candidate sites, and finally to a preferred site or sites. Typically, a Geographic Information System (GIS) is utilized. The appropriate region is identified, which might be a utility service area, a region, or even a state. The first screening is done down to candidate areas. This is in reality an elimination process, in which places inappropriate for a power plant are eliminated. These eliminated areas can include:

- pristine natural areas such as aquatic preserves or wildlife refuges,
- flood-prone areas,
- areas not zoned and not easily rezoned for industrial use,
- areas geotechnically unsuited for foundations,
- areas too remote from rail, heavy duty highway, and/or water-borne access,
- areas too remote from existing transmission lines,
- areas too remote from natural gas pipelines, and
- areas with no water supply.

The best way to perform a siting study that will satisfy the regulators is to involve outside environmental groups throughout the process. The groups that would be most likely to oppose the project later on should be identified up front and included in the site selection process. This includes not only national groups like the Audobon Society or the Sierra Club, but also local groups. If you are not familiar with these groups, it is worthwhile to hire specialists to help identify them. If you are a utility which has been opposed in the past in the identified region, you already know who they are. The most successful tactic has been to form a site selection task force or committee to either advise or perform the site selection study. Have committee members start with the list above and add or subtract items as they recommend. Make sure you put enough of your technical people either on the group or in consultation with it so that the technical siting requirements don't get left out.

Be sure to document what criteria you use, and which outside people you have consulted. 40 CFR 6.203 (d) includes a requirement to discuss "Coordination," which means who you have consulted with and what they said. This includes not only the federal, state, and local officials, but also any members of the general public who have been involved.

Candidate Sites

After you have identified candidate areas, you will need to pick actual sites to look at. Based on your plant conceptual design, you will be able to estimate how large a site you will need, what type of transportation facilities will be required, what size heat dissipation system will be required, and what power and voltage transmission characteristics will be required. Using this information, coupled with such economic factors as availability, you will need to identify actual sites capable of supporting the development of the desired project. This task is best done by site layout engineers, as there needs to be sufficient engineering to determine that each site is feasible. Once the candidate sites have been identified, a conceptual layout for each one should be prepared. This layout will be used to do the cost estimates for construction and operation that will be used in the evaluation. All of the candidate sites should be technically and environmentally feasible. The cost estimates must include any environmental mitigation that will be required. At this time, you should also be looking at the sites to answer the question, "If I have to show that this site will provide a significant environmental benefit to the environment, what would it be?" You will find that the only way to get the regulators and public in favor of your project is to show that it provides some such environmental benefit. The benefit could be wetlands enhancement or creation, or preservation of habitat for endangered species, or something else.

There have been numerous methodologies proposed to evaluate sites and rank them preferentially. These methods all involve some sort of weighted scoring system to do a total numeric estimate of the value of each site. Evaluation should include both environmental (or licensing) and cost-based criteria. Environmental criteria that have been used are included in the following areas:

- air quality,
- land use planning and socioeconomics,
- terrestrial ecology,
- aquatic ecology,
- hydrology/water quality and waste management, and
- geology and seismology.

Air quality criteria should include background air quality concentrations, impacts of plant emissions with respect to PSD increments, (see Chapter 5, Air Emissions and Controls, for a definition of PSD) future ambient pollutant levels with respect to ambient standards, impacts of plant emissions on nonattainment areas, and potential impacts of heat dissipation system operation.

Socioeconomic and land use considerations include aesthetic value (impact of the plant on the surrounding viewshed), employment, historic/archaeological values (effects on resources on or near the site), land use compatibility, property value of the site, public services (plant impacts on schools, medical facilities, and public safety), recreation, taxes, and transportation.

Terrestrial ecology criteria should include locations of endangered or threatened species, locations of other unique species or habitats, occurrence of valuable wetlands on site, occurrence of prime farmland on site, relationship of site to surrounding area, value of site as game habitat, and habitat diversity of the site.

Aquatic ecology criteria should include ecological value (recreation and commercial factors), sensitivity (habitat, community, and species diversity), health (ecological disruption or degradation), assimilative capacity (size and water availability), fatal flaw (vulnerable critical habitats), and special or unique habitat such as nursery or spawning grounds.

The evaluation of hydrological, water quality, and waste management characteristics of each site should include analysis of water availability, thermal assimilative capacity of the receiving body of water (RBW), makeup water quality, RBW quality, relative soil impermeability, relative nearness of groundwater, potential impacts to existing local water uses, and flooding potential and impacts.

Geological/seismological considerations are primarily related to engineering feasibility of the site. They should include acceleration level (seismic factor), liquefaction potential, foundation stability of subsurface materials, subsurface carbonates (potential solution cavities), and dewatering potential (groundwater characteristics).

Unless you have a technical specialty in one of these areas, it is advisable to use experienced consultants. They will be able to determine which of these considerations can be addressed briefly and which require more detailed analysis. When they have finished the job, make sure that the results are well documented in a complete and stand-alone report. A significant amount of time may elapse between the site study and selection and the proposed construction at the selected site. If the site study is well documented, it can be used at that later time without the necessity of hiring all those experts again.

Chapter 3
Baseline Site Characterization

The chief goal of regulators is to gage the impacts that a project will have on the environment. This is based on the requirements described in 40 CFR 1502.15 Affected Environment, which is in the CEQ rules for an EIS. Whether the project is a new power plant, or additions to an existing power plant, the permitting process requires that a comprehensive effort be expended to define in great detail the baseline against which impacts will be measured. For a new project the baseline may be forests and farmland; for an established plant undergoing upgrade or additions, the baseline will be the existing environment as it has already adjusted to the existing plant. Delineating the baseline involves researching all aspects of the site and its surroundings: the land, the water, the atmosphere, the plants, the animals, and the people.

Each of these areas will require in-depth study and the diversity of skills needed to do this makes baseline characterization a bigger challenge than most people expect. This is where the "ologists" come in. You will need a geologist, two hydrologists (one for surface water and one for ground water), two biologists (one for terrestrial ecology and one for aquatic ecology) and a host of other specialists to do a good job here. Regulators generally request data for a certain period of record. The "ologists" will know where to check the literature to find available data others have obtained. This can reduce the extent of field surveys to merely verifying that the literature data is representative of your site.

Whether you have the "ologists" in-house or contract the work out to consultants, make sure the job is done in enough detail the first time so the regulators won't come back asking for more. It is tempting to cut corners during the baseline assessment phase, but that won't save time or money if it delays the project. A common mistake is to ignore some aspect of the baseline because you don't believe your project will affect it. Later, you discover you have to change your project in some way that requires analysis of what you omitted. An example is a power plant project that involved a fuel change, with an associated increase in capacity factor. The existing plant had a permit to withdraw surface water sufficient to operate at full capacity, but actual plant operation had historically run much lower. Because the plant had been forced to stop making groundwater withdrawals, the utility assumed they would have to obtain the additional water for cooling system makeup to accommodate the higher capacity factor from the surface water. Deep in the licensing process, they were forced to agree to a water supply scenario that required use of significant quantities of groundwater and reclaimed water (recycled sewage effluent) as well as surface water. Unfortunately, they had not documented the groundwater quality or water levels and so had to take the time during the permitting process to obtain that baseline data. This resulted in a project delay of several months.

So, where do you start? Start with the land.

THE LAND

There are several questions that have to be answered about the land—how much is there, where is it, and what is it made of? The first two are best rolled into delineation, and the third is, of course, geology. You will have to describe the existing land features in sufficient detail so that later efforts at impact analysis have sufficient baseline against which to compare.

DELINEATION

Draw the site to scale on topographic maps with a scale of at least 1:24,000. Show clearly the outline of the proposed site and indicate whether the property is already owned by the applicant or whether there is any surrounding property which may be purchased to add to it. Label nearby landmarks, roads, railroads, etc., so they are obvious. Identify and label surrounding land owners and/or uses (e.g., pasture or industrial) on the maps also. Provide a breakdown of the total site acreage and its composition. If the land is presently undeveloped, give a description of its uses.

This first set of maps highlights the site and its immediately surrounding lands. Equally as important is giving a regional view of its location. Regulators want to know how the site fits into the big picture. Prepare more maps on a smaller scale showing where the site is located in relation to nearby towns and cities within a five mile radius. In fact, a series of five maps showing the surroundings within one, two, three, four, and five mile radii is preferable.

Identify any lands under governmental protection or with special environmental, historical, archaeological, or other, significance within a five mile radius of the site. A list of special lands to look for is given in Table 3-1. An archaeological field study may be required if one has not been performed before. Send a letter to the state historic preservation officer requesting a determination. Stereoscopic examination of aerial photographs can also prove useful in locating archaeological finds. Any large significant land holdings within that same radius in private hands should also be reported. Environmental groups like Greenpeace and the Nature Conservancy may own property which is used mainly as an environmental resource.

You will later have to present arguments that show your project will have little or no impact on these features. If you overlook one, a regulator or intervenor may well bring it up. The net result will be delays in the project while you and they evaluate potential impacts to the omitted feature.

Determine the government entity which has jurisdiction over the area in which the site is located and obtain copies of local zoning or land use plans from the city council, board of county commissioners or local planning

Table 3–1 List of Special Lands

Parks
Forests
Seashores
Wildlife Refuges
Wilderness Areas
Memorials
Monuments
Marine Sanctuaries
Estuarine Sanctuaries
Critical Habitat of Endangered Species
Roadless Area Review and Evaluation Areas
Wild and Scenic Rivers
Archaeological Landmarks
Conservation and Recreation Lands
Game Management Areas
Aquatic Preserves
Indian Reservations
Military Lands
Sites placed on the National Register of Historic Places

commission. Regulators prefer that no changes in zoning or land use plans be required prior to construction of the project. Investigate and describe any expected trends or changes in surrounding land use. Proof of title to the land in question should also be clarified. Keep in mind that any titles, easements, or crossing approvals required will be obtained from local governmental agencies.

Order a copy of the Flood Insurance Rate Map for the site area from the Federal Emergency Management Agency (FEMA). This map will designate whether the site is in a floodway, a flood plain, or neither, and to what elevation the local 100-year flood will rise. Because power plants are water use intensive, they are almost always located in or near a floodway or a floodplain.

If you have purchased the site, you will want an environmental assessment performed for protection against liability for previous owners' potential hazardous material mishandling. If you have not yet purchased the property, be sure to include at least a Phase I Environmental Site Assessment (ESA) to maintain your status as an "innocent landowner." This is particularly important if you are going to use a site that has already been put to industrial use, (e.g., a "brown-field" site). If you already own the property and have done so for a long time (e.g., a repowering project), you should probably do an ESA anyway, because you are likely to be digging up areas with potential for having been the locations of previously legal activities which left material behind that would now require remediation. You don't want the unexpected cost of such remediation to make your project suddenly uneconomical.

GEOLOGY

Once the surface questions have been answered, delve deeper into the land. This will be important for design of foundations for your project structures and probably for predicting impacts to groundwater. Contact the United States Geological Survey (USGS) or the state Geological Survey office for published literature describing the general geologic features of the site area. (Studies are often available by county of interest.) This description should include the following information:

- soils,
- underlying rock strata and its thickness,
- age (era, period, and epoch),
- aquifer types and zones,
- stratigraphy (formation and group), and
- formation or group composition and color.

Assemble a regional geologic map compiling this information and a separate map of the surface lithology and structure contours.

Have your geologist prepare a summarization of the geologic and soil studies performed during the site selection stage. The level of detail in this report depends on the complexity of the plant components to be built. For instance, if there will be no cooling pond, waste ponds, or coal pile, then the report is much simplified. However, you will most likely be doing soil borings for foundation design.

Conduct a baseline survey to develop the geologic cross section of the site. This is done via boreholes; make sure you use your foundation design borings as part of your survey. Ensure that proper borehole logs are kept in case they are requested by the regulators. Logs should include the location and approximate elevation, size, and depth of the borehole as well as drilling data such as date, driller, method, and rate. Once the borehole is abandoned, the closure method should be added to the logs as well as a description of the material used to fill the borehole. The geologist should prepare a lithologic profile from the rock cores giving depth profiles, grain size, porosity, hydraulic conductivity, and cation exchange capacity.

You will also need the soil classifications of the soils on site. Conduct soil borings to identify these, as well as to aid in assessing the bearing strength of the soils. The latter is important in determining the suitability of the site for supporting foundations for structures or material stockpiles. The soil classifications should be verified against Soil Conservation Service (SCS) soil maps.

The Water

This is a broad category. First, you want to know as much as possible about the available water resources. Second, you want to know who's using water from these resources and how much they are using. In other words, how much is left for you. Interestingly, one of the users of surface water is the aquatic life within the water, such as the fish. Increasingly, power plants are being asked to make use of other user's discharges, primarily that from municipal sewage treatment plants.

Water Resources

Water resources naturally fall into two categories: groundwater and surface water (although there may be interaction between the two). Either of these resources may serve as a source of water for the plant, and will certainly be considered under the impact phase of the permitting process. The baseline conditions for water resources should be described in sufficient detail for the agency to analyze the impacts of your intake and discharge upon the resources, whether you propose any or not (i.e., even for a zero discharge site).

Investigating the groundwater resources will require study of the subsurface hydrology of the site area. This includes circulation, flow, transmissivity, and water quality. Use a groundwater hydrologist to characterize the distribution and circulation of the groundwater system in the site vicinity. For an existing plant, this information may be available from previous investigations, particularly if an Environmental Impact Statement (EIS) or equivalent permitting document was prepared. For a new site, drill extraction wells and perform pump tests in order to characterize the hydraulic properties of the groundwater. These properties include the transmissivity, storage coefficient, and potentiometric surface. Determine any seasonal differences in water table contours, and amount of recharge, including areas and rates.

The use of groundwater has an advantage over that of surface water in that no one has yet found aquatic life in groundwater that would be impacted by its usage. However, if the groundwater is of good quality, other anthropogenic uses will compete with the power plant. Use of groundwater for drinking water is generally perceived as a higher or more valuable use than use for a power plant.

For groundwater quality data, conduct an initial sampling event, and based on the analytical results, prepare a monitoring well sampling program. Important parameters include the following:

- ionic concentration,
- hardness,
- total dissolved solids (TDS),

- pH, and
- temperature.

Hydrostatic head is also important to know to analyze mixing within multi-layered aquifers.

Surface water hydrology is the study of the dominant surface water features in the area. The surface water hydrologist should determine water level fluctuations, evaporative losses, flow, and quality for the water bodies of interest. Ascertain the drainage basin areas which are feeding each water body, taking into account seasonal variations, and obtain USGS records on river and stream flow and lake water levels in the vicinity. In addition, there may be some significant interaction between the ground and surface systems via hydraulic connections. This is particularly important for plants utilizing cooling ponds because seepage may affect both groundwater and surface water levels in the area. You will have to delineate the interface at which the groundwater and surface water hydrologists separate the work.

Hydrometeorological data (such as precipitation) can be ordered from the National Oceanic and Atmospheric Administration (NOAA) which maintains meteorological stations throughout the country. Coordination between the surface water hydrologist and the meteorologist will be necessary to avoid duplication of effort here.

Learn the regulatory classifications of the applicable surface water bodies. These classifications change regularly, and discharge permits depend heavily on them for determining discharge limits. All states classify the waters by usage (e.g., recreation, fish propagation, potable, etc.) because federal law requires them to do so. Special classifications are applied to waters which have extraordinary uses. One example would be the sole breeding location for a species of concern. However, a water body can receive a nondegradable classification merely because active environmental groups pursue it.

Hydraulic parameters that will need to be examined include the morphometry (shape of the water body), bathymetry (depth ranges), volumetric capacity, and stages (water levels). Stage versus capacity curves for each significant water body will need to be prepared. Reference the stage to mean sea level (MSL) and to National Geodetic Vertical Datum (NGVD). Different agencies with differing yet overlapping jurisdictions can each require their own reference level. MSL is a reference to the actual average of the nearest tidal level for at least 19 years. NGVD refers to a set of monuments which were installed and surveyed back in 1929. Although the two references are indistinguishable for most purposes at most locations, there are local areas at which they differ significantly. Also, collect any information on direction and magnitude of currents, wave heights, and tidal influence.

Conduct a water quality sampling program to measure temperatures and the levels of metals, nutrients, ions, and other parameters in the surface water. Generally, historical water sampling data can be used, if supplemented by a site-specific field program. The less historical data available, the more comprehensive the field program will have to be.

The selection of constituents is usually a blend of those pollutants expected from a power plant, and those for which water quality standards have been set. The list of those expected from a power plant is based on the worst case, a coal-fired power plant. This list is presented in 40 CFR 423, Appendix A. Those constituents for which water quality standards have been set can be determined in most cases by a review of the appropriate state water quality regulations.

In addition, you will want to obtain data for constituents which are either important for plant operation, or which are used to calculate the water quality limits. Parameters of interest include the following:

- alkalinity,
- hardness,
- total dissolved solids (TDS),
- dissolved oxygen (DO),
- total suspended solids (TSS),
- turbidity,
- pH,
- oil and grease,

- specific conductivity,
- dissolved metals,
- sulfates,
- chlorides, and
- silicates.

Ideally, site specific values for at least 12 consecutive months should be measured and compared to historical values for the same time period to verify the applicability of the historical values. If an analysis is performed based on unrepresentative data, it can cause permit violations later.

Plots of maximum, minimum, and average concentrations over time will need to be prepared. These can indicate any seasonal effects on concentrations. Special circumstances affecting surface water quality should also be investigated. For example, irrigation runoff could introduce a high nutrient concentration.

For an existing plant, surface water bodies already on site, such as cooling or stormwater retention ponds, require this same level of analysis, since impacts will be gaged against the existing conditions at the site.

Water Users

Once you have explored all of the water resources, you need a general accounting of who is using the water and how much they are using. Contact the agency which regulates local water use and/or operates the well permitting program. Identify any permitted wells or surface water withdrawal locations within a five mile radius of the site. Locate these on a vicinity map. Note that permits may list the maximum allotted usage or the maximum capacity of the well or surface water facility in question, rather than the actual usage.

Wastewater Dischargers

You will also need an accounting of who is discharging industrial wastewaters to the surface waters in the vicinity, especially if there is a possibility that their discharge could commingle with yours. You could be asked whether the two discharges could have a harmful effect together that is greater than the sum of their individual effects (this is called synergy). You could also be asked to use their discharge as your intake water.

The Atmosphere

This category is a combination of climate and air quality since these two items are interrelated. For example, when atmospheric pollutants lower the pH of precipitation, acid rain is the result. The first step in characterizing the baseline atmosphere is to research the meteorology affecting your site.

Meteorology

Locate the nearest National Weather Service (NWS) observation stations to your site. Stations are scattered throughout the country, primarily at airports. First-order stations collect a wide variety of data, including precipitation, ambient temperature, humidity, wind speed, wind direction, barometric pressure, and atmospheric stability. This last parameter is also called sigma theta. Obtain as much historical data as possible which is representative of your site, particularly if you have no on-site meteorological monitoring stations. You will be required to present air quality modeling which utilizes the full range of data from a first-order station. However, you may be able to use data from a non-first-order station rather than having to install an on-site meteorological station.

Representative data include daily and monthly temperatures, precipitation, relative humidity, barometric pressure, and seasonal and annual averages of dispersion characteristics, including hourly wind speeds, directions, and stability categories; mixing heights; and winds aloft. Present the ambient temperature means and extremes—daily and monthly—in tabular form. The variation between daytime and nighttime temperatures and humidity levels should also be described. Determine when fog is likely based on historical records. Give a description of rainfall patterns (i.e. what months of the year experience the most and the least).

Ascertain the acidic level of precipitation and surface water pH levels and the extent to which they have been affected by acidic precipitation. From studies, obtain annual rural deposition of laboratory H^+, excess SO_4^{2-}, and NO_3^- and concentrations of particulate SO_4^{-2}, SO_2, HNO_3, and NO_2.

Determine the prevailing wind direction at the site and how it varies throughout the year. If there is a large water body nearby from which winds originate, determine when they are most prevalent. Obtain wind data as available from the National Weather Service. These data should be made into wind rose figures which show the speed and direction on a frequency basis (See Figure 3–1 for a typical wind rose).

From historical records, characterize the frequency of severe weather events in the area (thunderstorms, high winds, heavy rain, hail, lightning, tornadoes, and hurricanes).

The atmospheric stability will be important in gaging the potential for pollutant dispersal. It is usually characterized as unstable, neutral, or stable, and these terms are related to temperature changes with height, which in turn affects diffusion rates. The literature may have classifications based on wind speed, cloud cover, and solar radiation reaching the earth's surface. Mixing height is another parameter which is important to determining the dispersion capacity of the atmosphere. It is dependent upon surface temperatures and solar radiation. Characterize the mixing heights for the site from empirical data from a source (probably an airport) and show as isopleths for time of day (e.g., differences between morning and afternoon).

Ambient Air Quality

You will need to provide at least one year's worth of ambient air quality data as part of the air quality program. Historical data collected by the state may be available to satisfy some or all of this requirement. If not, you will need to set up your own monitoring program. Follow EPA guidelines for prevention of significant deterioration (PSD) to locate monitoring stations and to choose the sample parameters and methods. The monitoring

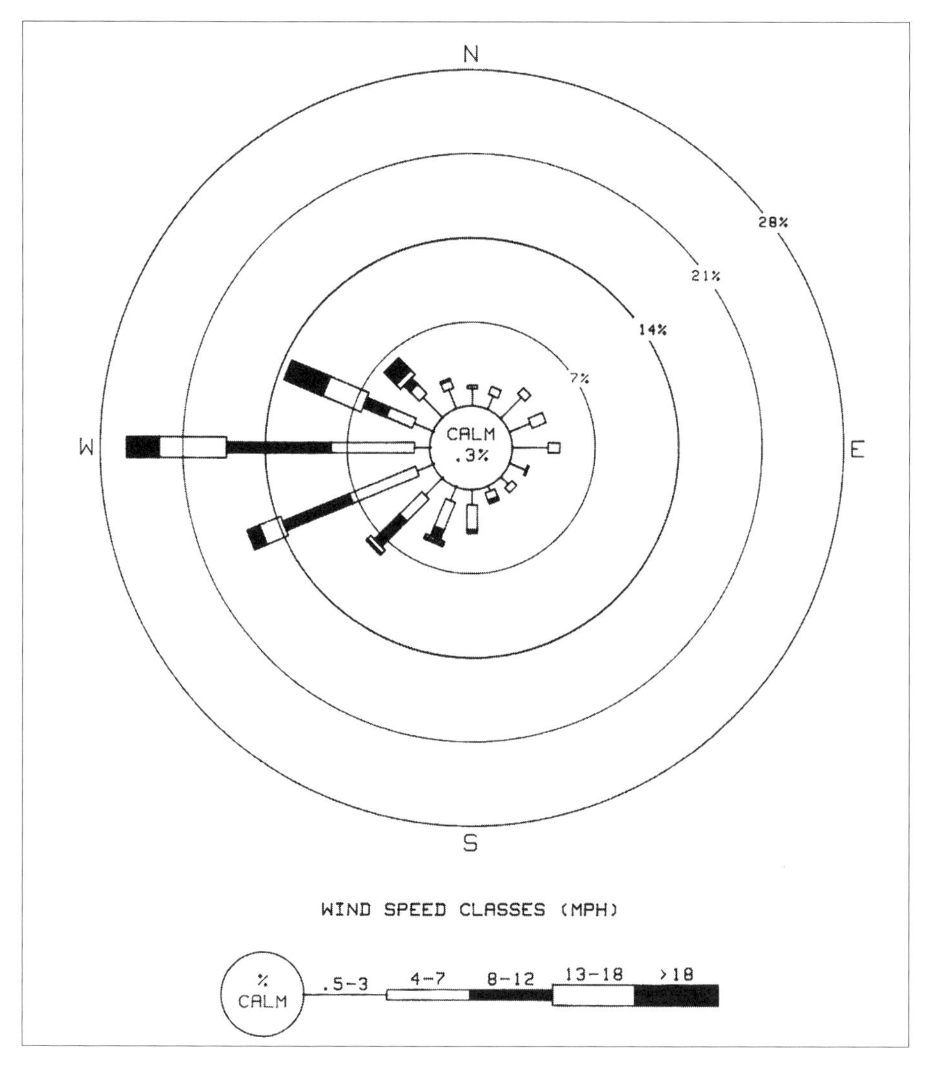

FIGURE 3-1 Typical Wind Rose

program will require approval prior to its commencement and should incorporate quality assurance procedures.

Parameters of interest are:

- sulfur dioxide (SO_2),
- nitrogen dioxide (NO_2),

- particulate matter (PM or PM_{10}),
- carbon monoxide (CO),
- volatile organic compounds (VOCs), and
- ozone (O_3).

An air quality monitoring program will need a professional to carry it out in order to be acceptable to the regulators. The state regulators for air quality are always susceptible to EPA review of their efforts, and are thus generally very rigorous. The air quality technician should keep detailed records about how the program is carried out, the locations and elevations of monitoring stations, types of monitoring instruments, and the frequency and duration of measurements. You will have to compare the data obtained to the National Ambient Air Quality Standards (NAAQS) and the air quality standards of the state.

It is possible that you may substitute modeling results for empirical data. The modeling program would require prior approval of the regulators who will want to see detailed background information concerning the validity and accuracy of any models to be used. You should not attempt to do this modeling yourself, unless you have taken and passed the EPA Air Dispersion Modeling course.

THE PLANTS

To start with, contact local regulatory agencies for a general vegetation map of the site area showing the major vegetation types such as range land, forest, wetlands, pasture, and prairie. Additional resources for information of this type include the United States Natural Resource Conservation Service (SCS, formerly Soil Conservation Service).

A detailed survey by a biologist (terrestrial) or botanist will be required to categorize the presence, abundance, and condition of vegetation in more detail. Such a survey would include a detailed list of species surveyed, the approximate acreage occupied by each habitat and its condition. Items of concern include the extent of the vegetation (acreage) and a qualitative assessment of the types present. It is important to note if the existing vegetation has been disturbed either by human acts (e.g., controlled burning or cattle grazing) and/or by natural causes (e.g., electrical storms, flooding, or wildlife).

At the present time, the most valuable vegetation is perceived to be the wetland species. There are federal, state, and local agencies that each have their own methodology of determining whether an area is a wetland or not. They all include the vegetation species as part of that determination, along with type of soil and the frequency and magnitude of occurrence of water. It is vital that a specialist who is qualified by as many of the appropriate agencies as possible be employed at this stage. The lead federal agency is the Army Corps of Engineers (COE). Specialists can be certified by the COE to do wetland jurisdictional determinations. State agencies are trying to become more consistent with federal criteria, but regional and local agencies may not see such a need yet. The specialist you hire needs to be familiar with the requirements of all the agencies, including the local ones. Therefore, local experience is mandatory.

Also of major importance will be the presence of any species considered threatened, endangered, or under some other type of regulatory protection. Lists of endangered or threatened flora are usually compiled by state agencies.

The Animals

With apologies to the biologists of the world, we group aquatic life forms under this category even though they don't all belong in the strictest scientific sense. The first step is to find scientists to traipse around your site and vicinity, and to identify all of the animals living or visiting there. Perhaps you can use the same terrestrial scientist who traipses around looking at the vegetation. Two separate efforts will most likely be needed: terrestrial ecology and aquatic ecology.

Terrestrial Ecology

The chief goal of this survey is to characterize the wildlife populations of the existing site in enough detail so that the regulators can be sure the project will not harm or disturb species of importance. Species of importance are listed by the U.S. Fish and Wildlife Service (FWS) and other agencies, such as the state fish and wildlife service and state game and fresh water fish commission. In addition, species which are unique or endemic to the area or which are dominant in the habitat may fall under this category. Game and fur-bearing animal populations should be identified, as well as whether the existing site is used for hunting. Birds are included as members of the terrestrial group. The survey should also examine habitat quality in general and any existing stresses on the wildlife. Such stresses can include: domestic animal grazing, feral hog disturbances, fire damage, and human activity. These stresses may have already altered the composition and abundance of the species in the region.

Aquatic Ecology

Aquatic ecology "animals" include organisms from fish to plankton as part of an interrelated ecosystem. An unbroken and healthy food chain is key to a thriving system. Your specialists should perform sampling to characterize the aquatic systems present in the vicinity of the site. A series of surveys performed during different seasons would garner the most useful and reliable information. As with terrestrial ecology, the species of importance are determined from regulatory agency lists.

Survey results should include a map of the sampling locations and a cross-sectional view of the streams sampled. The survey should include examination of habitat conditions, such as bank vegetation, the littoral zone (area between high and low water marks), substrate types (stream bottom sediments), current direction and velocity, air temperature, water temperature, pH, DO, conductivity, salinity, and light transmissivity of the water. The water quality constituents of these data can also be used in the surface water quality description.

Unless the information is available in the literature, your specialists should also perform a fisheries survey to determine what species inhabit the area surrounding the site and what part they play in the overall ecosystem in the area. Survey results should be presented in a table with the scientific and common name of the fish species and number collected. For multiple surveys, a percentage breakdown of the species observed is a useful way to present information. The presence of migratory species should also be noted. Spawning and breeding is evidenced by the presence of nests with eggs or if juvenile fish are captured during the survey. If neither is present then the area probably is not a significant breeding area. Of particular importance during this survey is the presence of any species under regulatory protection. These can range from species which are endangered or threatened to those which are considered game fish.

Another useful bit of information among fisheries people is called the Fulton coefficient of condition (K). K values are a means by which scientists compare the relative physical condition of different fish species. K values are published in the literature and provide a range of favorable values to which to compare.

You will also need to conduct a benthic macroinvertebrate survey. Benthic macroinvertebrates are bottom-dwelling creatures such as clams or snails which are a significant source of food for fish. The density and diversity of benthic macroinvertebrates present is a good indicator of the overall healthiness of the ecosystem. Survey results should include the scientific name of the organisms captured, their numbers, and their density.

Don't underestimate the importance of these organisms. A power plant in the Northeast had never had discharge effluent limits because the river it discharged to was considered lifeless due to acid mine drainage. During an NPDES permit renewal, the state agency decided the river had recovered enough to impose discharge limits because a few benthic worms were observed under a bridge. The net result was to require the plant to install a new wastewater treatment plant.

THE PEOPLE

Baseline environment, as defined in power plant permitting, is not just the physical, biological, and ecological resources on and around a site, but also that slice of humanity which will experience any side effects, good or bad, from the project. There are sociological ripples from constructing and operating a major facility. Regulators will request information that seems to you to be unrelated to the project—the local school system, for instance. They will use this information in models to determine what kind of impacts the project will have on the local community.

Impacts will be measured on the labor force, housing market, public services, and even the aesthetic quality of the existing site (noise and visual qualities). The scope of information required for this baseline can be very wide and surprisingly detailed. This task usually requires the services of a consultant with community planning experience. This is an area in which a new or improved power plant can be expected to have a major positive impact. Therefore, it is important to have a consultant with a proven track record.

The first step is to delineate the socioeconomic extension area of the project. The extension area could include only the local city or county or a larger area. It is usually defined by determining where the majority of the affected population lives. For an existing plant, this can be measured from a study of the commuting patterns of the employees. The larger you can make this area, the more likely it will be that the negative impacts (such as increased traffic) are less significant, and that the positive impacts will be more significant. It is important to show positive impacts (e.g., increased jobs and income, as well as increased spending in retail establishments) in the local area, so as to garner local support.

Detailed data on the local populace is catalogued through a demographic profile including statistics on population, density, labor force, wages, and housing. Information regarding public services, utilities and transportation capacity availability is also presented. It is best to present most of this information in tabular form. A list of suggested tables is given in Table 3-

2. The consultant should present future trends also, such as expected population growth rates or proposed major transportation projects.

Resources for some of this socio-political data include: local planning agencies; U.S. Bureau of the Census; local community growth and development agencies; the state Board of Labor; economic, business, and civic groups; the state and local Boards of Education; the State Department of Commerce; and local sheriff's and fire departments.

TABLE 3–2 Socioeconomic Tables

Population Projections
Labor Force and Unemployment Statistics
Estimated and Projected Employment by Profession
Employment by Industrial Sector
Per Capita Income
Average Annual Wages of Workers Covered by Unemployment Compensation
Number of Households by Income
Housing Characteristics
Building Activity
Public Education Systems
Major Transportation Facilities
Levels of Service
Public Service Facilities
Potable Water Facilities
Sewage Treatment Facilities
Solid Waste Facilities

You will also need a noise expert to perform a noise survey to assess that portion of the aesthetic status of the existing site. This survey should be performed under the direction of a qualified noise specialist and done to EPA guidelines. Part of the survey effort is to locate any noise-sensitive receptors within the site vicinity. These include residences, schools, churches, hospitals, nursing homes, and public parks. Traffic, railroad, and aviation noise would also be considered in assessing the baseline. There are no federal noise standards, but the EPA does have guidelines for noise levels. You must determine if your state or locality has any noise ordinances which will affect your site. Many local governments do.

Other

There may be other environmental features of your site which have not been discussed above but which need to be examined as part of the baseline. These features should be discussed with the appropriate regulators to determine what course of study should be followed. In particular, if your site has a present environmental problem (e.g., it was formerly a strip mine or dump), it is important to establish that problem with the regulators, with the intent to incorporate a solution to that problem within your project.

Chapter 4
Site Specific Design Alternatives

Chapter 2 explained the regulatory basis behind the requirement to do an analysis of alternatives, and went on to discuss the alternatives, which could be analyzed without having a specific site selected. These included alternative means of satisfying the need for the project, alternative sites analyses, and the no-action alternative. Once you have picked a particular site, the analyses of remaining alternatives will be set against the background of that site. These site specific design alternatives are the subject of this chapter.

Table 4–1 identifies eight specific areas for which alternatives need to be addressed by your engineering staff to provide you with a feasible and economical design for your power plant. Include an environmental analysis of these systems in with the engineering analysis for two main reasons. The first reason is that a regulator can ask you at any time to justify the design you are proposing. If you are asked, and permitting has become the critical path for your project, you will be much better off if you can pull a study off the shelf and show that environmental concerns were integral to your decision-making process.

The second reason is that a significant cost factor could be omitted from the engineering study if the environmental factors are ignored or not included properly, thus invalidating the results. We have a good example of just such an occurrence. It happened on a repowering project in which new combustion turbines were being added. A study was performed to determine whether the combustion turbines should be inside an enclosure

Table 4–1 Site Specific Design Areas

Site Specific Design Areas
Condenser Cooling System
Plant Water Systems
Wastewater Treatment System
Air Pollution Control System
Solids Management
Transmission Lines
Plant Location/Layout
Fuel Handling System

or not. The initial results of the study indicated that the cost of the enclosure could not be justified on strict engineering grounds (e.g., less down time due to easier maintenance).

Then, the environmental conditions were factored in. Our client found that they would have to install a noise wall around the combustion turbines to meet the local noise ordinance and a segregated drainage system to prevent potentially oily stormwater runoff from being discharged. On the other hand, if the enclosure were built, the roof drainage would be of good enough quality to use as a water source for one of the plant water systems. In this case, the environmental factors justified the costs of what initially appeared to be an expensive and unjustified enclosure.

We will examine each design area in Table 4–1 and list alternatives from our experiences that you should consider.

Condenser Cooling System

The condenser cooling system, or heat dissipation system (which includes other minor cooling systems which are generally negligible in comparison to condenser cooling), is typically such a large system that it is treated as a separate system from the other plant water and wastewater systems.

To put the size of the systems into perspective, the piping for the condenser cooling system typically ranges in diameter from five to as much as 16 feet (see Figure 4–1). The only non-power facilities using such large pipes are typically big-city sewage treatment systems, stormwater systems, or water supply systems. Other industrial facilities, and other power plant water systems, typically use piping ranging from mere inches to three to four feet in diameter.

In a typical fossil-fuel fired power plant, up to 10 percent of the fuel heat content is wasted up the stack. Between 35 and 60 percent comes out as electricity (60 percent for a CC unit, 35 percent for a conventional steam electric unit). This leaves between 30 and 55 percent of the fuel heat content as waste heat lost to the condenser cooling system. In a typical nuclear plant, it is even worse. There is no heat up the stack and the electrical power coming out is typically about 33 percent of the fuel heat content. Thus 67 percent of the heat value coming into the plant leaves as waste heat lost to the condenser cooling system.

The condenser transfers the heat from the steam to the circulating water, raising its temperature by as much as 30 degrees Fahrenheit. There are several methods of dissipating that waste heat. The oldest power plants had once-through cooling systems that transferred the heat to the nearest large water body. As implied by the name, the circulating water was withdrawn from the natural water body, passed through the condenser once, and was then discharged back to the same water body. The discharge location was selected so that the heated water would not be recirculated back into the system again. The natural water body was in effect a heat dissipation device. Heat dissipation systems evolved to match increasingly stringent environmental restrictions.

The alternative to the use of once-through or open cycle systems is to use closed cycle systems. In these, instead of a natural water body, an artificial heat dissipation device is used. Most of these dissipate the heat to the atmosphere.

The type of heat dissipation system for your site should have been selected during your site selection studies (see Chapter 2). Now that you

FIGURE 4-1 Cooling Water Pump Impeller

have a site, you must do a more rigorous analysis of all the alternative heat dissipation devices available.

These devices include:

- dry cooling towers,
- wet-dry cooling towers,
- cooling ponds,
- cooling sprays (see Figure 4–2), and
- evaporative cooling towers.

FIGURE 4-2 Cooling Sprays

If you are at an existing power plant site, you may be able to take advantage of an existing once-through system or cooling pond. In most locations, however, you will likely conclude that an evaporative cooling tower is the best choice. In that case, you will have to look at all the different types:

- natural draft (see Figure 4–3),
- fan-assisted natural draft,
- round mechanical draft, and
- rectangular mechanical draft (see Figure 4–4).

FIGURE 4-3 Natural Draft Cooling Towers

WATER SOURCE

As part of your heat dissipation study, you will have to select a water source. Even closed cycle systems usually need makeup water to replace evaporation. To ensure smooth plant operation, you will want to use the

FIGURE 4-4 Rectangular Mechanical Draft Cooling Tower

highest quality water you can find. Regulators will want you to use the worst quality water they can find. This is because water is high in demand and low in supply. This trend shows no signs of reversing in the near future.

Your study should examine the following water sources:

- groundwater,
- surface water (rivers, lakes and reservoirs, estuaries, or oceans), and
- reclaimed water.

Reclaimed water is defined in various ways across the country. In this book, we mean water that has been used and is a wastewater and is then treated sufficiently to be reused. In many areas the term is meant to specifically refer to treated domestic wastewater (sewage treatment plant effluent).

Intake System

If your water source is a surface water body, study alternatives for your makeup water intake system. A good discussion of the appropriate methodology is in the EPA publication entitled "Guidance For Evaluating The Adverse Impact Of Cooling Water Intake Structures On The Aquatic Environment: Section 316(B) P.L. 92-500," published May 1, 1977, by the Industrial Permits Branch of the Permits Division.

The different alternatives to consider are:

- location with respect to shoreline (inlet flush with shoreline, offshore inlet, or open canal to inlet),
- location with respect to depth, and
- location with respect to balance of plant and discharge.

Discharge System

You may have a discharge system for cooling system blowdown (see Figure 4–5) or for once-through cooling if you are repowering an existing unit. Your discharge could be to a receiving body of water (RBW) or to another user as reclaimed water. Your choices for RBW include surface or groundwater. Analyze these alternatives and also alternative locations within the RBW (i.e., surface or subsurface and with respect to the intake location).

A good reference for this evaluation is the EPA document entitled "Interagency Guidance Manual and Guide for Thermal Effects Sections of Nuclear Facilities Environmental Impact Statements," published May 1, 1977, by the Industrial Permits Branch of the Permits Division.

You may decide that all of this work might not be able to get you a permit, even for a discharge from a closed cycle system, and that your best choice is a zero discharge system. In a zero discharge system, you take the blowdown from your closed cycle system and treat it so you can reuse it. Even with this option, you will have material that you have to get rid of, but it will be a solid waste instead of a liquid one.

FIGURE 4-5 Blowdown Discharge

Methodology

Table 4–2 suggests a stepwise methodology to go through when doing your alternative condenser cooling system analysis.

Plant Water Systems

The condenser cooling system has already been discussed as a heat dissipation device. However, it is also a plant water system, so analyze it with respect to biocide usage and physico-chemical treatment alternatives (anti-scaling or anti-corrosion additives). Biocide alternatives include gaseous chlorine, sodium hypochlorite, bromine chloride, chlorine dioxide, ozonation, and ultraviolet radiation. Chances are that some form of chlorine will be the engineering choice, due to the biocidal effects of the chlorine residual. In order to avoid the requirements for a Risk Management Plan under the accidental release prevention program under Title III of the Clean Air Act, and the potential for a disastrous accidental release, avoid the use of gaseous chlorine.

The preferred biocide will most likely be sodium hypochlorite.

Table 4–2 Alternative Condenser Cooling System Analysis Methodology

1. Identify each source and its water availability.
2. Identify each RBW and its thermal assimilative capacity.
3. Determine if once-through cooling can meet thermal and biological (impingement and entrainment) criteria.
4. Select once-through or closed cycle (with or without discharge).
5. Perform a conceptual layout of the selected system, including intake and discharge, and of at least 2 other possible systems.
6. Perform an economic and environmental comparison.

There are several other potential plant water systems. If you have a wet flue gas desulfurization (FGD) system, it will require makeup water. FGD systems are very forgiving; they can physically accept almost any water as makeup, including demineralizer regeneration wastewater. Water source quality limits are generally only imposed if the byproduct (scrubber sludge such as gypsum) is to be recycled to another industry as a raw material (e.g., wallboard or cement industry).

If you have any steam electric capacity, you will have steam generator (boiler) makeup that must meet very high purity requirements. Boiler makeup usually requires both pre-treatment and advanced treatment. Pretreatment alternatives include sedimentation, filtration, softening, and carbon absorption. Advanced treatment alternatives include ion exchange, reverse osmosis, electro-dialysis-reversal, and evaporation.

If your plant has a combustion turbine (e.g., a CC plant, or a simple cycle CT), you may have an inlet air cooling system, and it could be of the evaporative type. These typically require good quality water so as to avoid scaling from minerals left behind when the water evaporates.

Service water system is the name usually given to a combined system serving various small miscellaneous water needs such as pump bearing cooling, pump seals, air preheater washing, equipment maintenance cleaning, and other miscellaneous plant uses. The service water system often serves as the source to the potable water system, if the plant does not use city water, and to the fire protection system. In such a case, the service water treatment system can be thought of as pretreatment for the potable and fire protection water systems. Treatment alternatives for service and/or potable and fire protection water include:

- screening,
- sedimentation,
- filtration,
- disinfection, and
- softening.

If your plant is an oil or coal-fired unit, it will most likely have an ash handling system. Although fly ash is preferentially handled dry, bottom ash and pyrites (from coal-fired plants) are better handled wet, as the water can be used to seal the boiler against air intrusion from below. Although older plants often have once-through ash systems, these have generally been replaced with recirculating systems to minimize the discharge of pollutants.

A coal-fired plant will often have a wet dust suppression system to minimize the generation of fugitive dust emissions from the coal pile or coal handling equipment such as conveyors or stacker/reclaimers (see Figure 4–6).

Figure 4-6 Coal Bucket-Wheel Stacker-Reclaimer

The methodology for performing the alternatives analysis for the plant water systems is tabulated in Table 4–3.

Wastewater Treatment System

Most of the plant water systems give rise to wastewaters that require treatment before they can be discharged or reused. Depending on your site,

TABLE 4–3 Alternative Plant Water System Analysis Methodology

1. Evaluate the particular system demand requirements for both flow rate and for physical/chemical quality.
2. Evaluate potential source water alternatives including groundwater, surface water, and reclaimed water (both plant and off-site water).
3. Analyze treatment requirements for each source.
4. Prepare a conceptual design for each feasible alternative source.
5. Perform an engineering and environmental comparative analysis.

you may want to treat each of them locally in a separate treatment facility, or you may want to collect them all into one central wastewater treatment system. You may want to do a little of both. Interestingly, the EPA and state regulators have a bias against central wastewater treatment facilities because they can't measure the quality of a particular effluent stream by itself, before and after treatment. This means that they can't tell if you are meeting effluent limitation guidelines (40 CFR 423). They call the mixing of wastewater streams "commingling." It is a classic case of the letter of the law being put above the intent of the law.

The major plant process water systems that give rise to wastewater streams include:

- condenser cooling system,
- steam generator,
- FGD system,
- ash handling system,
- water treatment system,
- wastewater treatment system,
- radwaste system (nuclear plants only), and
- inlet air cooling system.

In addition to these process-generated wastewaters, there can be several other wastewater streams at your power plant. These include rainfall runoff streams from:

- material handling areas,
- material storage areas,
- material disposal areas,
- equipment areas (potentially oil-contaminated), and
- non-contaminated yard areas.

Possible materials are fuel, limestone and other additives, and waste products such as ash or scrubber sludge.

Another source of wastewater is maintenance cleaning. This includes boiler cleaning on either the fireside or the air side, by chemical or non-chemical means. If you are using acid for cleaning, use organic acids so the wastewater won't meet the RCRA definition of hazardous wastewater (i.e., pH less than 2). Wastewaters also come from cleaning equipment, cleaning the condenser, cleaning the cooling tower(s), and even just washing the floors.

Another wastewater found at all power plants, but not specific to them, is sanitary wastewater (domestic sewage). It is generated in bathrooms, shower rooms, and break areas. Rural power plants typically have their own package sewage treatment plant, mostly of the extended aeration type (see Figure 4–7). The effluent is usually much smaller in flow magnitude than other effluent streams, and is discharged into one of them. It is not uncommon for it to be recycled into the heat dissipation system. Connecting to a municipal system is generally a more cost-effective option, if it is available. This is because the cost of having a licensed operator for your sewage treatment plant, and of doing all of the required monitoring, generally exceeds the cost of paying the municipal system to take the wastewater stream. Also, most municipalities and comprehensive land use plans favor hooking into municipal systems.

Just as there were miscellaneous water uses, there are miscellaneous wastewater streams. These include the same items mentioned above with respect to service water such as pump bearing and seal water. Also included in miscellaneous wastewaters are the drains from your plant water quality laboratory. As with demineralizer regeneration wastes and metal cleaning

FIGURE 4-7 Sewage Treatment Plant

wastes, you will have to be careful to avoid having to handle the lab wastes as hazardous waste.

The methodology for analyzing the wastewater treatment alternatives is summarized in Table 4–4.

AIR POLLUTION CONTROL SYSTEM

REQUIRED CONTROL LEVELS

Nuclear units (except for older units with fossil fuel-fired auxiliary boilers) typically do not need an air pollution control system. Although they often have emergency diesel generators, the operating hours for these

Table 4–4 Water Treatment Alternatives Analysis Methodology

1. Identify all waste streams.
2. Identify flow rates and water quality for each waste stream.
3. Determine the suitability for reuse for each waste stream, with and without treatment.
4. Determine treatment requirements for remaining discharges to meet regulatory requirements, either separately or in combination.
5. Perform a conceptual design of each feasible alternative treatment system.
6. Perform a comparative engineering and environmental analysis of each alternative.

devices are so low that no air emission permitting is required. If you are permitting a fossil fuel-fired unit of significant size, you will have to do an alternatives analysis for the Air Pollution Control System (often called the Air Quality Control System). Base your alternatives analysis on the PSD (Prevention of Significant Deterioration) analysis as described in 40 CFR 52.21(b)12 (see Chapter 8, Permits). This is actually the definition of Best Available Control Technology (BACT).

Although technically you are not required to perform a PSD control technology review for pollutants for which your area is nonattainment (meaning that ambient air quality standards are not met), the same type of review will identify for you what is the Lowest Achievable Emission Rate (LAER) for that pollutant. If you are in an attainment area, PSD review is required for every pollutant subject to regulation under the Clean Air Act for which a "significant emission rate" has been defined that your plant will exceed. Significant emission rates as of this writing are summarized in Table 4–5.

Obviously, the best option for you is to design your project so that you do not exceed significant emission rates for anything. This is difficult to achieve unless you retire some other air emission source, preferably an older unit on the same site.

TABLE 4–5 Significant Emission Rates[1]

Pollutant	PSD Significant Emission Rate (TPY)	Applicable Standard(s)
Carbon Monoxide	100	NAAQS[2], NSPS[4]
Nitrogen Oxides	40	NAAQS, NSPS
Sulfur Dioxide	40	NAAQS, NSPS
Particulate Matter (PM_{10})	15	NAAQS
Total Suspended Particulates (TSP)	25	NAAQS, NSPS
Volatile Organic Compounds	40	NAAQS, NSPS
Lead	0.6	NAAQS
Asbestos	0.007	NESHAP[3]
Beryllium	0.0004	NESHAP
Mercury	0.1	NESHAP
Vinyl Chloride	1	NESHAP
Total Fluorides	3	NSPS
Sulfuric Acid Mist	7	NSPS
Hydrogen Sulfide	10	NSPS
Total Reduced Sulfur	10	NSPS
Benzene	Any	NESHAP
Inorganic Arsenic	Any	NESHAP
Radionuclides	Any	NESHAP

1. Compiled April, 1996
2. NAAQS = National Ambient Air Quality Standards
3. NESHAP = National Emission Standards for Hazardous Air Pollutants
4. NSPS = New Source Performance Standards

The pollutants that you will have to describe treatment and required control levels for generally will include the following:

- particulates from combustion sources,
- particulates from fugitive emissions,
- sulfur dioxide,
- nitrogen oxides,
- carbon monoxide,
- hydrocarbons (volatile organic compounds or VOCs), and
- hazardous air pollutants (HAPs) such as lead, beryllium, mercury, and fluorides.

Treatment Methods

Treatment methods for particulates emitted up the stack (predominantly fly ash) include electrostatic precipitators (ESP), fabric filters (baghouses), and wet scrubbers.

Significant fugitive emissions sources include coal and limestone handling systems, fly ash and FGD waste handling systems, unpaved roads and cleared areas, and cooling towers (drift). Controls for the first four include enclosures for piles and conveying systems, dry dust collection systems (fabric filters), and water spray dust suppression. Cleared areas are controlled with revegetation and cooling towers are fitted with drift eliminators. Generally, you will need either an ESP or a baghouse unless your fuel is natural gas.

Sulfur dioxide emission levels can be controlled by recovery or throw-away systems (known as FGD or scrubber systems). Both types of systems include dry, semi-dry, and wet systems. Wet systems use either a slurry or a clear liquor. The other way to control sulfur dioxide is to limit the amount of sulfur in the fuel. Natural gas and oil sulfur levels can be readily controlled by the supplier. Coal sulfur levels can be reduced by coal cleaning processes. If your fuel is natural gas, low sulfur fuel oil, or low sulfur coal, you should not need a scrubber. If you are using high sulfur fuel oil or coal, or other high sulfur fuel such as Orimulsion, you should expect to need a scrubber. This is true even if you have a fluidized-bed combustor which utilizes a limestone or dolomite bed to remove some of the sulfur dioxide.

Nitrogen oxide (NO_x) emissions come from two sources, the nitrogen in the fuel and the nitrogen in the combustion air. The most efficient control of NO_x emissions that are caused by thermal oxidation of nitrogen in the combustion air is by good furnace design, including furnace size, burner spacing and design, and excess air. Once the NO_x is produced, the available back-end treatment methods include selective catalytic reduction (SCR) and selective non-catalytic reduction (SNCR).

Carbon monoxide and VOC emission control methods are similar to NO_x control methods. They include furnace design on the front end, and catalytic reduction on the back end. In general, the best way to control these emissions is with good furnace design. You should not have to add back-end (catalytic reduction) treatment to control these emissions.

Control methods for lead and beryllium emissions are generally the same as for stack particulate emissions (i.e., either baghouses, ESPs, or scrubbers). You should not have to add any extra treatment for these. Similarly, fluoride emissions are generally as hydrogen fluoride (HF), and are controlled by whatever FGD system you have. If you don't have one, your fluoride control should be fuel content. Particulate mercury control is similar to that for lead and beryllium. Mercury vapor control is more problematic. EPA has been looking at incinerator controls as being applicable for other combustion sources. Control methods being proposed for incinerators include injection of sodium sulfide or activated carbon upstream of the particulate control device, and the use of wet scrubbers. You should expect to have to do a separate study on mercury BACT unless your fuel is natural gas.

Solids Management

Solids produced by a power plant have traditionally been called solid wastes. However, since the advent of RCRA, the preferred method of disposing of this material is to recycle it as a raw material to some other industry. Since the material is then not a waste, it is usually called a byproduct. Therefore the term "solids" as used in this section refers to both solid wastes and to solid byproducts. The unique solid waste from a nuclear plant is the used fuel rods. These require special handling under specific safety rules, because of the radioactivity involved, and are not addressed herein.

Virtually all steam electric plants produce water and wastewater treatment solids. These are generally disposed of to a landfill (either on site or off site). They are difficult to market, but could be used as a combination fertilizer/soil amendment. Another solid waste produced at virtually all power plants is refuse. This includes such material as office wastes, circulating water screenings, and combustion turbine air inlet filters. This material is typically disposed off site to a landfill as there are no practical alternatives.

Coal and other solid fuel-fired plants produce material handling solids, such as coal and limestone dust. These materials are usually best recycled within the plant, but can be landfilled.

Solid and liquid fuel-fired plants produce, to varying degrees, combustion byproducts. Bottom ash can be wet sluiced to an ash pond, or to dewatering equipment for subsequent landfill disposal or reuse. Bottom ash can also be mechanically removed for landfill or reuse (see Figure 4–8). Fly ash can be wet sluiced to an ash pond, or it can be pneumatically conveyed (dry) for on-site landfill, off-site disposal, or reuse (e.g., as an ingredient in the manufacture of cement). Pyrites (unique to solid fuels) can be handled in the same ways as bottom ash. FGD wastes can also be handled wet (pond disposal) or dry (landfill disposal or recycled). There are numerous off site reuse possibilities for this material, including gypsum as a wallboard ingredient and ammonium sulfate as a fertilizer.

FIGURE 4-8 Bottom Ash as Road Fill

Table 4–6 addresses the methodology to use in performing solid handling alternatives analysis for non-hazardous solids. It is prudent to show an on-site capability to dispose of all material for which you do not have firm contracts for off-site disposal, even if you do not intend to use it.

Anything that you generate that falls into the category of hazardous waste will have to be addressed separately. Design alternatives to minimize the generation of hazardous wastes should include the following:

- avoidance of water and wastewater treatment chemicals that render resultant sludges hazardous due to high pH,
- neutralization of demineralizer regeneration wastes in a TETF (totally enclosed treatment facility) or an ENU (elementary neutralization unit),

- recycling of boiler blowdown containing hydrazine,
- use of non-hazardous metal cleaning solutions such as citric acid, and
- accumulating miscellaneous hazardous wastes (e.g., toxic laboratory wastes and wastes from painting and degreasing operations) in proper containers which are stored in a central storage-for-disposal area for less than 90 days before off-site disposal to a licensed hazardous waste facility, according to the management standards of 40 CFR 262.34(a).

TABLE 4–6 Solid Handling Alternatives Analysis Methodology

1. Identify expected volumes and expected characteristics.
2. For on site disposal, identify location and size of site, and any liner requirements.
3. For off site disposal, determine contractor availability and willing landfill facility.
4. For reuse, perform a marketing survey and negotiate a contract if possible.

TRANSMISSION LINES

Begin your alternatives analysis for installing new transmission lines (or even for upgrading existing ones) with system stability studies to indicate where the power should go. Perform engineering feasibility analyses to determine operating voltages and currents. Once you have identified the beginning and ending points of your transmission line(s) for the new power, the study becomes strictly environmental.

The next step is to select alternative corridors from which to choose. The process is essentially one of exclusion. You want to maximize the amount of existing right-of-way that is used, hopefully minimizing the amount of new right-of-way that is required. You will have to avoid the following as much as possible:

- wetlands,
- floodplains,
- important habitats,
- historical and archaeological sites,
- sensitive land use areas,
- endangered species,
- important aesthetic areas,
- open water bodies, and
- infringing on the public convenience.

Once you have avoided all of the above that you can, what is left should allow you to select a preferred corridor. You can justify this corridor using both an environmental and engineering/cost comparison. For example, you might shorten your route by spanning a water body, but increase your costs by having to use helicopters to avoid building an access road.

After selecting the preferred corridor, you will probably be able to obtain a conditional permit that allows you to build the line by following conditions in selecting the actual right-of-way. Conditions might include adjusting your pole separations to avoid localized sensitive areas. Finally, you will need to evaluate alternative clearing methods including herbicide, burning, and manual clearing.

Plant Location/Layout

You will need to address the following plant layout alternatives:

- number of generating units,
- size of generating units,
- type of condenser cooling system,
- type of air quality control system,
- anticipated solid waste volumes, and
- type of fuel handling system.

You will also need to address how you evaluated the following facility location criteria:

- geotechnical/foundation requirements,
- minimization of earthworks,
- flood protection,
- wetlands avoidance,
- separation of stack and cooling tower plumes (if applicable),
- avoidance of drift on sensitive locations (if applicable),
- land use criteria,
- noise criteria,
- aesthetics criteria,
- avoidance of surface and groundwater contamination, and
- minimization of rail and road construction.

Fuel Handling System

You already performed an alternative fuel analysis prior to selecting your site. Your fuel should include one or more of the following types:

- natural gas,
- uranium,
- fuel oil (or Orimulsion), and
- coal (or other solid fuel).

If your fuel is natural gas, your delivery will most likely be via a pipeline. You should let your gas supplier license and build the pipeline right to your site. Typically, the on-site portion of the pipeline will be yours. Considerations of pressure and capacity will be dictated by your generating equipment on one end, and by the gas company's pressure, capacity, and pipeline material. Your air permit will most likely require you to meter fuel delivery rates for each unit. If you have liquefied natural gas (LNG) delivered to your site, you will have to contend with FERC regulations and design criteria, as a matter of public safety. Avoid this if you can.

If your fuel is oil (or Orimulsion), as either primary or backup, you will probably have aboveground storage tanks. Delivery to your site can be by pipeline, truck, ship (tanker or barge), or even by rail. These delivery methods should have been selected in your site selection studies (see Chapter 2). The questions you will have to address for on-site alternatives include number and size of tanks, areas of below versus aboveground piping, type of secondary containment (see Figure 4–9), minimization of production of potentially oily stormwater runoff (see Figure 4–10), and segregation of it from other stormwater runoff.

FIGURE 4-9 Concrete Secondary Containment for Aboveground Oil Tank

If your fuel is nuclear (uranium), your fuel handling and delivery will be under the scrutiny of the Nuclear Regulatory Commission (NRC), and you will not be allowed very much flexibility in the ways you deliver and handle your fuel.

If your fuel is coal (or other solid combustible fuel such as wood or wastes), the delivery could be by rail, truck, or water-borne (ship or barge). For rail delivery, you will have to evaluate unloading options including trestle, bottom dump pit, or rotary dumper unloading. Active storage options include reclaim by rotary plow feeder or by bucket-wheel stacker-reclaimer (see Figure 4–6). If delivery is by truck, you will need a method of weighing each load (see Figure 4–11).

If you have a water-borne coal delivery system, your choices for unloading include continuous bucket ladder barge unloader, a receiving bin for a self-unloading barge/ship, or a barge/ship unloader. You will need a plant delivery system either by overland conveyor (see Figure 4–12), rail, or truck.

Figure 4-10 Roof Preventing Runoff from Oil Transfer Equipment

Figure 4-11 Coal Truck Scales

Figure 4-12 Overland Conveyor System

Chapter 5
The Plant

Regulators cannot analyze the possible impacts of your project without knowing the details of what comprises your project. They are required to do an analysis by 40 CFR 1502.16, the Environmental Consequences portion of the CEQ rules for preparation of an EIS. Section 1502.16 requires that they consider direct and indirect effects to the environment, possible conflicts between the proposed action and any existing land use plans, and the means to mitigate adverse environmental impacts, if there are any. Their informational needs include the layout, the number and size of units, fuel, water intake and discharge systems, material handling, waste handling systems (solid, hazardous, sanitary, wastewater), stormwater management system, air emission control devices, and any other applicable systems.

In our experience, this kind of information is usually retained in parts by various segments of the engineering team, which may or may not be communicating adequately with each other. Since you know the regulators are going to request all of this information, it is best to compile it into a stand-alone report which can be used throughout the permitting endeavor as the definitive source. See Table 5–1 for a suggested outline for the report. Each of the topics will be discussed in detail below with hints for achieving sufficient coverage of all of them.

TABLE 5–1 Outline for Plant Report

- I. General Information
- II. Generating Technology
- III. Fuel and Fuel Handling
- IV. Air Emissions and Controls
  - A. Emission Sources
  - B. Proposed Control Technologies
- V. Water Usage
  - A. Heat Dissipation
  - B. Potable Water
  - C. Process Water
- VI. Wastewaters
  - A. Process
  - B. Sanitary
- VII. Solid and Hazardous Wastes
- VIII. Site Drainage
- IX. Materials Handling

GENERAL INFORMATION

Start the report with a good general description of the project's physical features. Include the acreage of the site, the types of units and fuel(s) to be used, and brief descriptions of major air, water and waste systems. For an existing plant, describe how the new project will interact with the existing components. For instance, there may be shared facilities, like the cooling system. Include a large-scale layout diagram showing the site perimeter and the location of the various facilities within it. The footprint of the property should be recognizable on the topographic maps that you prepared earlier for the site description (see Chapter 3).

Include an artist's conception or architectural rendering (see Figures 5–1 and 5–2) of the plant showing its approximate appearance after

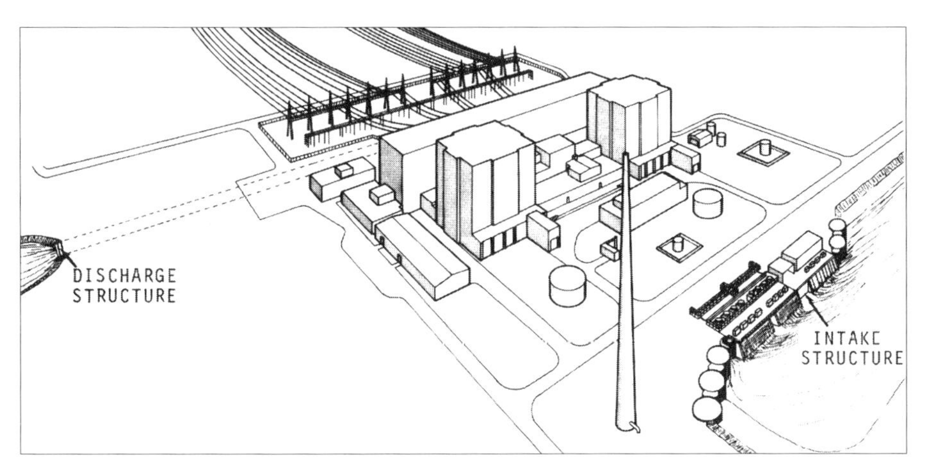

Figure 5-1 Artist's Conception of the Plant

Figure 5-2 Architectural Rendering of the Plant

construction is completed. This will be utilized later in a viewshed analysis to assess aesthetic impacts of the plant.

Generating Technology

As described in Chapter 2, the majority of plants being built today are utilizing fossil fuels either in a steam electric or combustion turbine (CT) configuration. CTs can be either simple cycle (no steam generation) or combined cycle (CC). The latter method is preferred due to cost and fuel efficiency considerations. Most of the descriptions which follow are based on CC units or coal-fired generating technology. Many new plants are being proposed to burn waste fuels such as tires, wood wastes, and even bagasse. The coal-fired option is representative of these waste fuel-fired choices.

For this section of your report, prepare a process flow diagram which represents the process in boxes and arrows. Electric generating plants are unusual in that none of the raw materials that go in to the plant are found as a tangible component of the product. They are, instead, all converted to byproducts. This is because of the intangible nature of the product electricity.

In addition, present small-scaled layout plans, detailed footprints of the units and cross-sectional views. These plans and cross-sections need to be at least 11 inches by 17 inches in size. The lateral location of gaseous and liquid waste release points should be marked on the footprints and their elevations listed on the cross-sectional views. This locational information will be utilized later as input to your predictive analysis for impact assessment.

Briefly describe the associated facilities which are connected to the generating system. These might include receiving, storage, and preparation areas for fuel and feed materials (e.g., limestone for desulfurization systems, lime for water treatment system); conveyance systems (e.g., conveyors for coal or limestone, or pipes and pumps for fuel oil or natural gas); oxygen generating plant (such as for a fuel gasification facility); byproduct handling/storage areas and any other auxiliary equipment areas which will be utilized.

Provide brief justifications for your system choices (e.g., fuel choice was based on economics and transportation considerations.)

Fuel and Fuel Handling

Most plants have at least dual fuel capability, or redundant methods of supply and delivery if there is only one fuel. Include information about the type, quantity, quality, distribution, handling, and storage of the fuels. Provide a condensed version of the design considerations which influenced your fuel choices and what alternative fuel types were considered and dismissed (e.g., waste derived fuel). Also describe what fuels were considered and are proposed as backup and/or future fuels. If the plant site was located to make use of a particular coal mine or gas field (a so-called "minemouth" plant), explain the justification for this choice (usually economic).

Discuss availability of the fuels and present typical analyses of their important characteristics. For natural gas, fuel oil, and coal, analyses should include ultimate and proximate gravimetric breakdowns of the fuels. An ultimate analysis gives the percentage of each element present (carbon, hydrogen, oxygen, nitrogen, and sulfur). A proximate analysis gives the percentage breakdown of the following components: volatile matter, fixed carbon, moisture, ash, and sulfur. For fuels obtained from various providers (typically coal or oil), present a range for the analyses values. If your plant will use coal gas, give a range for the volumes by percent expected of the following constituents: CO, CO_2, H_2, CH_4, N_2+Ar and H_2S+COS (carbonyl oxysulfide). Also, list the heat of combustion or higher heating value expected.

All of these data will be used by your experts (or the regulators if you don't have your own experts) to predict the efficiency of your units, the quantity of associated waste heat which will be transferred to the environment, and the amounts of solid byproducts and gaseous combustion byproducts which will be produced and transferred to the environment.

Discuss your proposed fuel delivery methods, usage rates for each fuel, and storage/handling considerations. This information will be used to analyze the potential for fuel spills or leaks during delivery and handling, and the environmental impact of such occurrences. It also is relevant to analysis of impacts on the existing transportation systems, for example, due to increased truck, train, or ship traffic.

For a coal-fired plant, describe the groundwater protection methods you will utilize on the site. These will probably include a coal pile liner and a leachate collection and removal system. Stormwater and drainage will be discussed in more depth later in this chapter.

Air Emissions and Controls

Because of the inherent complexities of air permitting, the discussion of air emissions and controls at your plant will constitute a major effort. First, you must determine (or have someone determine) whether you are expected to be a major source of air pollutants, as defined in 40 CFR 70. Unless you are retiring and replacing older units, you will almost certainly be a major source. If you are also in an attainment area (defined in 40 CFR 81, Subpart c), then you must summarize your Prevention of Significant Deterioration (PSD) application, which is the groundwork for determining the Best Available Control Technology (BACT). If you are in a nonattainment area, you will have to demonstrate that your emissions controls meet Lowest Achievable Emission Rates (LAER). For more details, see Chapter 8.

Begin by describing what emission sources you will have, the proposed control technologies for handling them, and expected emission rates. Your description of the controls will include a lengthy regulatory analysis because emission control selection is driven by regulatory considerations.

EMISSION SOURCES

Perform an emissions inventory to quantify the emissions expected from each source. Emission rates are best estimated from the manufacturer's data for point sources whenever possible. An EPA document entitled "Compilation of Air Pollution Emission Factors, Volume I: Stationary Point and Area Sources, AP-42," (commonly referred to as AP-42), presents methodology to estimate emission rates for area sources and for point sources when no manufacturer's data are available. The plant will probably have primary sources of air emissions (combustion turbines or boilers) and numerous secondary emission sources, such as:

- material handling systems,
- auxiliary boilers,
- cooling towers,
- storage facilities,
- diesel generators,
- coal, limestone, or gypsum piles,
- flare stacks,
- incinerator stacks, and
- oxygen plant vents.

Discuss all emission sources and under what conditions they will operate (e.g., only during start-up and shutdown versus most of the time). Some sources are only applicable to emergency situations or are operated infrequently. Make note of any planned emission source retirements which can be used to offset the new emissions. You must determine what condition constitutes the worst case from an emissions standpoint. For example, worst-case, full-load CC operation will generate varying tons per year of pollutants depending on the fuel used. Sulfur and nitrogen oxides are usually lowest using natural gas and highest using fuel oil. Emissions from a coal-fired boiler will vary with the variability in the composition of the coal. Your choice for air emission controls must be based upon the expected air pollutant emissions during worst case conditions.

The pollutants of interest are as follows:

- sulfur dioxide (SO_2),
- nitrogen dioxide (NO_2),
- carbon monoxide (CO),
- volatile organic compounds (VOCs),
- particulates,
- lead,
- beryllium,
- mercury, and
- inorganic arsenic.

Significant emission rates for these (and other) pollutants are listed in 40 CFR 51.24, "Prevention of Significant Deterioration of Air Quality." The project is subject to PSD/BACT review for pollutants which exceed those criteria, if you are in an attainment area.

PROPOSED CONTROL TECHNOLOGIES

The purpose of the PSD application process is to determine what control technologies constitute BACT for each regulated pollutant listed in 40 CFR 51.24. This is a "top down" approach which takes into account economic, environmental, and efficiency concerns of the individual plant. Thus, BACT varies from plant to plant. However, no matter which technology is chosen as BACT for each emission source, the proposed emissions cannot exceed the applicable New Source Performance Standards (NSPS) of 40 CFR 60. Chapter 60 has numerous subparts which cover various sources of emissions. Subparts Da, Db, and Dc cover electric utility steam generating units of various sizes. Subpart GG covers stationary gas turbines. Subpart J, "Petroleum Refineries," can serve as guidance for gasification fueled plants since EPA has not yet promulgated standards for gasifiers.

Preparation of the PSD application and the corresponding BACT evaluation is a major effort which is performed interactively with the regulators. The evaluation includes consideration of various technologies and/or methods and the justification for those chosen as BACT. Have your experts do a thorough analysis to support your choice for BACT. The regulators will try to make you move to a higher level, and you will have to convince them that you are right in what you propose to do.

Include all applicable information in your plant report along with design parameters of the proposed control equipment. Your choice of control equipment will be dependent on generating technology and fuel characteristics. For instance, natural gas is a very clean burning fuel; its sulfur content is only a trace amount. Therefore, burning it "as-is" does not exceed NSPS, so no additional control is required. In such a case, BACT for SO_2 from natural gas fired CC units could be simply burning low sulfur natural gas. If the units are to also burn fuel oil, you might restrict the oil's sulfur content so that it could be considered a low-sulfur fuel also.

Common control technologies for controlling emissions in CC units are as follows:

SO_2	Use of low sulfur content fuel Flue gas desulfurization (FGD)
NO_x	Use of low nitrogen fuel Steam or water injection Dry combustion Selective catalytic reduction (SCR) Selective noncatalytic reduction (SNCR)
Particulates	Clean combustion Controlled maintenance Inlet air filtering Steam injection
CO and VOCs	Combustion control Oxidation catalysts

Similarly, for coal-fired units, common control technologies include:

SO_2 and acid gases (HF and H_2SO_4)	Use of low sulfur (and low fluorine) content fuel Wet scrubbers Spray dryer absorption Lime or limestone injection (fluidized bed boilers)
NO_x	Combustion controls (low excess air, flue gas recirculation, low NO_x burner design) Overfire air and/or gas cofiring Selective catalytic reduction (SCR) Selective noncatalytic reduction (SNCR)
Particulates	Clean combustion Controlled maintenance Wet control techniques (scrubbers, tray towers, wet ESPs) Electrostatic precipitators Baghouses (fabric filters)
CO and VOCs	Combustion control Oxidation catalysts

Water Usage

Perform an overall accounting of water sources and uses for the plant. Prepare water budget diagrams which show the flow quantities to and from all the plant systems for both average and peak conditions. Present quality data for the water sources to be used and the water quality required by the user systems. More and more, regulators are performing (or requiring) demonstrations that utilities intend to use the worst quality water possible.

Using potable water to generate electricity is looked upon unfavorably, unless there is just no other choice. Describe in detail each of the major water using systems at your plant: heat dissipation, potable supply, and process waters.

Heat Dissipation

Describe the heat dissipation method (e.g., cooling towers or cooling pond), how it will operate, and include a physical description of its layout. Subtopics for this section include:

- quantity of heat dissipated,
- quantity of water withdrawn,
- consumptive use,
- design size and location,
- blowdown volume and physical characteristics,
- temperature changes and retention times,
- evaporation rate,
- design plans for dams or dikes (see Figure 5–3) when a cooling

Figure 5-3 Cooling Pond Splitter Dikes

pond or storage reservoir is to be built (cross-sections, plan views, plot plans, and seepage rate),

- design and location of intake structures (water depth, flow and velocity, screens, trash disposal, number and capacity of pumps),
- maximum discharge temperature, and
- travel time from condenser inlet to discharge point.

For a cooling pond, provide the pond's surface area and storage volume, the maximum, normal and minimum operating levels, information on seepage and any underdrain system, where and when makeup is supplied, and a description of the intake structures. For a cooling tower, describe whether it is forced draft or induced draft (see Figure 5–4), cross flow or

FIGURE 5-4 Forced Draft (Foreground) and Induced Draft (Background) Cooling Towers

counter flow; mechanical or natural draft; hyperbolic, round, rectangular, or octagonal; and wet, wet/dry, or dry. For either cooling system, present the heat load expected in Btu/hr and the temperature rise across the condenser. Whether you choose a pond or towers, be prepared to defend your choice by showing that it will not use more water than its counterpart.

Describe any dilution system to be used.

Use computer modeling to predict the thermal performance of the pond or cooling towers. The model will rely on historical meteorological data, pond or tower design parameters, temperature differential across the condenser, and total heat load to be handled. Remember that for an existing pond the presence of aquatic growth or shoals can reduce the thermal efficiency of a pond by reducing its effective surface area. Other modeling input data include: precipitation, wind speed, relative humidity, percent sunshine, and air dry bulb temperature. Meteorological data can be obtained for long periods of record from the National Weather Service. Wind speed is not necessary for predicting the thermal performance of cooling towers.

Model output should include the expected average and maximum cold (condenser inlet) water temperatures and the expected evaporation losses, both natural and forced by the heat load. Predict these values on a monthly average and extreme basis. They will be used later during the impact assessment (see Chapter 7).

Use the evaporation estimates from the thermal modeling along with the intake or makeup water quality to predict the water quality in the cooling pond or tower. The blowdown from the heat dissipation device will be of the same quality. Use these estimates, known as "simple mass balance" estimates, to demonstrate compliance of your discharges with applicable water quality standards.

Include a discussion of disposal techniques for blowdown, screened organisms and trash collected from the intake structures.

Potable Water

Domestic (potable) water usage is normally limited to wash rooms, sinks, and water fountains. Indicate whether your supply will be obtained from public services or via treatment of on-site surface water or groundwater. Describe the quantity and quality of the water to be handled. If a treatment system is to be used, include a description of its design parameters and capabilities. Address any waste disposal issues associated with the system, such as backwash water from filters.

Most states have moved the regulation of potable water systems out of the jurisdiction of public health agencies and into that of environmental agencies. However, potable water systems are historically linked with public health concerns, and this linkage has produced a large body of regulations on these systems. They are very closely regulated. Thus, even though the capital cost for a potable water system is a very small percentage of the cost of the whole project, the amount of time and money that can be required for compliance will seem inordinately large. Minimize your effort by either connecting to a municipal potable water system (and letting them deal with compliance issues) or, if the project is on an already-developed site, reuse as much of the existing potable water system as possible. Hooking into a municipal system usually helps your project to meet land use goals and wins support from revenue grateful municipal officials.

Process Water

Describe the uses for process water in the plant, any treatment systems associated with it and the quantity and quality of pollutants to be discharged from the treatment systems.

Process water demands can typically include:

- boiler makeup water,
- fire protection,
- washdown water,

- dust suppression water (e.g., for coal, limestone, gypsum, fly ash),
- FGD scrubber makeup water,
- inlet air evaporative coolers makeup,
- laboratory uses, and
- miscellaneous plant needs (e.g., bearing cooling water and pump seal water).

Specify the users at your plant and their water demands on a water use flow diagram (see Figure 5–5). Present the quality characteristics of the raw water to be used and describe the required water storage facilities. For groundwater sources, include a map of the well locations and what aquifers they access. Address wastes generated by the treatment systems and plans for their disposal.

In general, you will want redundant sources of fire protection water for insurance and safety purposes. A cooling pond makes an excellent backup source; this should be factored into the selection process of a heat dissipation system.

Wastewaters

Prepare a complete characterization of the quantities of wastewaters that will be generated and the expected water quality of each stream. In the past, this information was used to design disposal systems to meet all applicable discharge criteria. The trend now is to examine each of these streams as a potential source for the water use requirements listed above. Recycling of wastewaters back into the plant reduces the overall quantity of wastewaters, making it easier (and hopefully cheaper) to treat and dispose of them. If you can reuse all of your wastewater, you will have a zero discharge system.

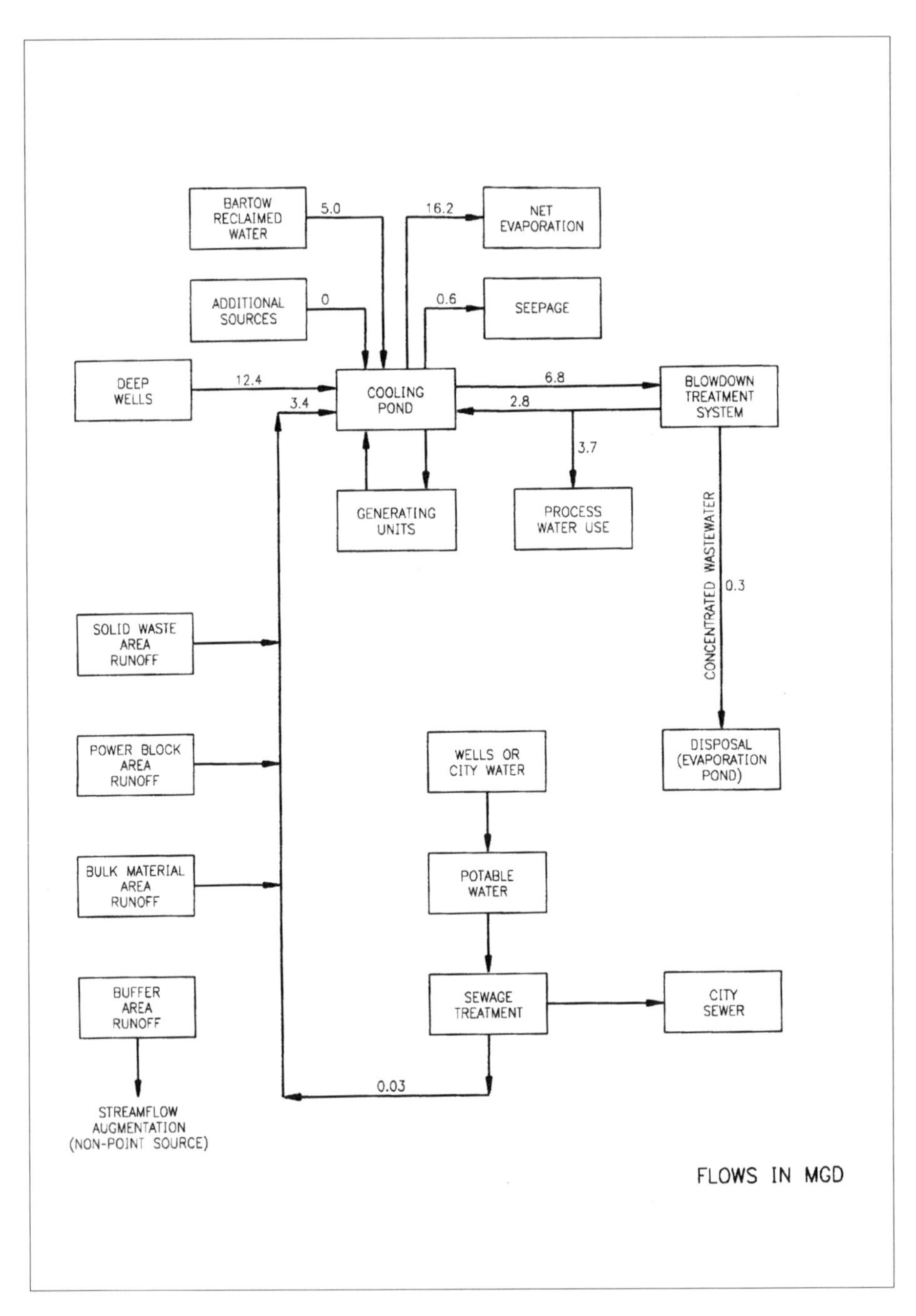

Figure 5-5 Water Use Flow Diagram

Process Wastewaters

Prepare flow diagrams for the chemical waste system, including descriptions of any chemical additives and the treatment methods planned. Include any blowdown discharge from the cooling pond or cooling towers. Chemical constituents will concentrate in these heat dissipation devices due to evaporation effects, thus discharges from them (including cooling pond seepage) will contain the same chemical constituents.

Other wastewaters include unit water treatment system wastes (e.g., demineralizer regeneration wastes, filter backwashes, reverse osmosis [RO] reject waters), cleaning and washdown wastes, boiler blowdown, equipment drains, wet FGD scrubber blowdown, laboratory and sampling wastes, oily wastes, chemical metal cleaning wastes, treatment chemicals, and other chemicals. Present detailed descriptions of the processes generating these wastes and the expected quantity and quality of them. Address any material handling wastes, such as coal pile or limestone pile storage runoff or leachate, or wastestreams from other material handling areas (e.g., ash or scrubber sludge disposal areas).

If injection wells are planned for disposal of boiler blowdown or other wastes, the regulators will want to see the well construction diagrams. Other requested information will include casing depth, well bore diameters, and grouting. Show the location of the monitoring wells on a map and give a description of the drilling and testing programs that are planned. You should have obtained detailed geological information on the planned injection well sites during your baseline assessment. In general, we recommend that injection wells be avoided due to the present regulatory bias against them and the cumbersome regulatory process required to license them.

Sanitary Wastewater

Sanitary wastewaters are discharges from bathrooms, wash houses, drinking fountains, eyewashes and other non-industrial sources. Indicate whether your plant will be hooked to the public sewer or will utilize an on-site treatment system. If a treatment system is to be used, include a description of its design parameters and capabilities. Describe the quantity and quality of the wastewaters to be generated. Parameters of particular concern are the BOD and TSS loading. Address any waste disposal issues associated with the system, such as sludge.

As with potable water systems, there are historical public health concerns about sanitary wastewaters. Likewise, the burden of regulatory compliance for them will appear much larger than their capital cost seems to justify. Most jurisdictions require that system operation be overseen by a licensed wastewater operator. Connecting to a municipal system (or your own existing system) will usually be preferable to installing your own new treatment works.

Solid and Hazardous Waste

Include in your plant report a description of all solid and hazardous wastes expected as a result of your project.

Solid Wastes

All solids produced at a power plant are wastes relative to the primary product, which is electricity. If you can find a use (either for yourself or for someone else), the waste becomes a byproduct. Itemize the solid byproducts and wastes to be generated at the plant, including their quantity and the methods for handling, storing, and disposing of them.

Solid wastes can include the following:

- general wastes such as office wastes, grass clippings and other yard wastes, and screenings from circulating and makeup water system screens,
- water and wastewater treatment wastes (filter backwash solids and water softening sludges),
- combustion turbine wastes (e.g., used air inlet filters),

- combustion byproducts such as fly ash, bottom ash, and pyrites,
- gasification wastes like slag and elemental sulfur and low volume wastes such as spent catalyst, solvents, and spent resin beds,
- FGD scrubber or spray dryer wastes (e.g., gypsum), and
- waste oils from oil/water separators or spill cleanup materials (see Figure 5–6).

FIGURE 5-6 Oil-Water Separator

Describe any liners to be used in the landfill areas. Also, discuss the chemical and physical composition of each of these wastes and how these characteristics affect the design of disposal facilities, and the possibilities for marketing the wastes as byproducts (i.e., recycling or reusing them off site in other industries). For example, fly ash can be reused as an ingredient in cement manufacture.

You may have good prospects for marketing some or all of these wastes. If you actually have contracts in place to do so, be very explicit about the quantities involved, how the wastes will be handled on site and transported off site. If you do not have signed contracts in place, present a plan for waste disposal, either in an on site landfill or off site. One creative disposal method which has been proposed is to ship coal ash back to the originating mine for reinjection.

HAZARDOUS WASTES

Hazardous wastes are defined under federal law. Specifically, they are characterized in Subpart C, and listed in Subpart D, of 40 CFR 261. State laws defining hazardous wastes may include other material not covered under the federal rules (e.g., certain petroleum products and, in some cases, combustion byproducts). However, the state rules will at a minimum include all of the federally defined materials. There are five categories of hazardous wastes listed:

1. non-specific sources,
2. specific sources and discarded commercial products,
3. off-specification species,
4. container residues, and
5. spill residues.

A power plant has the potential to generate wastes from all of these categories except the third. To minimize permitting hassles and cost, avoid storing any of them on site for more than 90 days and do not treat or dispose of them on site at all. We recommend you hire a hazardous waste contractor to dispose of your hazardous waste, and to keep track of all the paperwork required.

Determine which of your processes may generate hazardous waste, both listed and characteristic. If you will have gasification of coal or other feedstock, potentially hazardous waste sources include wastewater treatment wastes, acid gas removal wastes, sulfur recovery wastes (spent refractory), and tail gas treatment wastes. Gasifiers can also produce non-thermal wastes, which may contain hazardous chemicals.

Design your power plant so that you do not generate unnecessary hazardous wastes. We recommend that you avoid generating three types of hazardous wastes:

1. non-thermal wastewaters containing hazardous chemicals,
2. waste oils containing hazardous constituents, and
3. miscellaneous hazardous wastes (paints, solvents, toxic laboratory wastes).

You can avoid generating non-thermal wastewater hazardous wastes by avoiding treatment chemicals that require high pH levels in the wastewater (e.g., above 12). If you have a demineralizer, use a totally enclosed treatment facility (TETF, as defined in 40 CFR 260.10) or an elementary neutralization unit (ENU, also defined in 40 CFR 260.10) in order to be exempt from being regulated as a hazardous waste treatment unit under 40 CFR 265.1(c)(9). An even better strategy is to eliminate ion exchange units, and use RO and/or ultrafiltration instead, so that there will be no regenerants. Recycle boiler (steam generator) blowdown, which may contain hydrazine but is a very high quality wastewater, into another water system, like the heat dissipation system. Use non-hazardous chemical metal cleaning solutions, such as citric acid, instead of hard mineral acids for boiler cleaning. Have the spent acid taken off-site by a contractor for disposal.

Waste oil containing hazardous constituents could become a problem if you have a spill of fuel oil that contains biocides. Route drainage from all areas which could be affected to a central wastewater treatment system.

This will keep the levels of any biocides below hazardous levels.

Miscellaneous hazardous wastes could include paints, solvents, and toxic laboratory wastes. Prepare a hazardous waste management plan in which you discuss their minimization, storage, treatment and/or disposal. Although preparation of such a plan may not be specifically required in your state, it will help you minimize your hazardous waste generation (and associated costs) as well as help you show compliance with federal rules. This plan should include using containers of 55 gallons or less capacity in which these materials can be accumulated and a storage-for-disposal area (see Figure 5–7) where the containers can be stored for less than 90 days. If you meet the management standards listed in 40 CFR 262.34(a), you will not need a hazardous waste storage permit.

FIGURE 5-7 Storage-for-Disposal Area

Site Drainage

Site drainage from a power plant is subject to federal regulations described in 40 CFR 423, under the low volume waste category of the industrial NPDES permit program for any stormwater runoff potentially affected by power plant activity. This drainage is also subject to the stormwater NPDES permit program.

Many states also have stormwater regulatory programs. Some areas also have regional agencies which have stormwater management authority and finally, some also have local (county or municipal) agencies and rules. They all have the common goals of protecting people and installations from flooding and of protecting the water quality of receiving waters. Some have performance requirements, but many actually impose design requirements. The stormwater management system will require engineering, and the engineer(s) must ensure that the design meets all of the applicable agency requirements.

Describe the site drainage plans for your plant and locate all important features on a map. Include any detention ponds (see Figure 5–8) and discharge points for stormwater (See Figure 5–9). Identify the design storm which has been used by the engineers. Compare post-development discharge to existing discharge. Quantify the acreage of the drainage areas. Present a plan for handling drainage both during construction and operation, including planned control measures to prevent erosion. Map the locations of retention ponds, potential spoil areas and discharge points for stormwater runoff.

The same trend we described earlier for reusing and recycling wastewater is also occurring in stormwater management these days. Some facilities have been proposed and built that have zero stormwater discharges and recycle all of their stormwater. One potential problem with this approach is that the receiving water inflow is reduced, hence the flow in the receiving water is reduced, which may cause adverse impacts. Examples of such adverse impacts include increased temperature in the stream during summertime because of longer retention time and increased

Figure 5-8 Stormwater Detention Pond

Figure 5-9 Stormwater Discharge Point

saline intrusion in freshwater streams that empty into estuaries. Some jurisdictions now have limits on what percentage reduction of flow to a receiving water is allowable. Conversely, you could generate an increased inflow to the receiving water by increasing the runoff coefficient of the site due to industrialization. Ideally, your project could balance this increase by reusing some of the runoff.

Materials Handling

Discuss the construction materials expected to be unloaded, transported, and disposed of at the site. Describe how delivery will be made of large components and the locations for storage areas, laydown areas, and their pollution control features. Include a discussion of the transportation system around the site and its adequacy for transporting heavy equipment. Show construction facilities on a map. The system for handling operational materials such as fuels and sludges should also be discussed. Include lubrication oil handling; handling for gases such as hydrogen, oxygen, and nitrogen; and any other items not specifically discussed elsewhere.

Chapter 6
Impacts of Construction

The analysis of impacts is at the heart of the permitting process. The regulatory requirements for this analysis originate in 40 CFR 1502.16, Environmental Consequences. This portion of the CEQ rules relates to Environmental Impact Statements (EISs), calls for discussion of "any adverse environmental effects which cannot be avoided," and requires discussions of both direct and indirect effects, their significance, and whether they are irreversible and unavoidable. During construction of your project there will be impacts on the baseline environmental conditions defined earlier. You will have to provide regulators with information on those impacts so it can be determined whether they are small or large, whether they could be mitigated and how much opposition to the project will be precipitated by those impacts. This chapter deals with the impacts caused by the construction phase only. The impacts of operating the plant are dealt with in Chapter 7.

We suggest you compile an environmental management report on the construction phase of your project. This report should include a description of the expected impacts and their magnitude, their frequency of occurrence, their duration, and other applicable descriptive information. In addition, you should outline the steps you plan to take to monitor, test, sample, etc. for evidence of the impacts and what you plan to do to lessen or mitigate the impacts.

A great deal of this information will be required for some aspect of your construction permit application process anyway (e.g., Pollution Prevention Plan or Erosion and Sedimentation Control Plan). This includes a detailed description of what construction activities will take place; the temporary and permanent soil stabilization practices to be utilized; stormwater management; dust, odor, and noise control measures; clearing and burning methods; herbicide/pesticide application; and material handling and waste disposal.

Obviously a "green-field" site will be impacted more than an already developed site. It is best to stress throughout your report that construction activities will be conducted in such a way as to minimize environmental impacts. Begin with the resources you investigated and defined during the baseline assessment because they are the potential impactees of your construction activities. Describe the physical effects of site preparation and plant construction on these resources: the land, the water, the atmosphere, the plants, the animals and the people.

The Land

The first step is to outline the construction activities to take place and delineate the land (size and location) that will be affected by them. Next, describe the erosion control measures that will be utilized during construction. Finally, examine other aspects of construction that may have an impact on the land (e.g., blasting or open burning).

Construction Activities

Construction activities should be scheduled in a sequence which allows for proper controls to be utilized and for a minimum of land disturbance. Soil disturbing activities could include:

- clearing and grubbing;
- grading;
- excavation;
- backfilling;
- dewatering;
- driving piles;

- embankment construction;
- razing of existing structures;
- construction of basins, ditches, or ponds;
- installation of pond liners;
- building of access roads, rail siding, or barge unloading facilities;
- installation of temporary and/or permanent electrical and utility piping systems;
- construction of generating units and ancillary features; and
- construction of landscaping berms.

Present a general site map outlining the areas where these activities will take place. If any portion of the site will be set aside as a buffer or preserved area and will remain undisturbed, show its location on the map also. You will have to document the steps that will be taken to prevent construction from impacting such lands. This could include placing barricades around the area to be protected or some other marking system.

Include the areas that will be designated for construction vehicle parking, material storage, laydown, and disposal.

Provide the general sequence of construction activities which will take place. Include information on plans for stockpiling and reusing topsoil and the source to be used if additional fill material is required.

EROSION AND SEDIMENTATION CONTROLS

Once the sequence of activities has been outlined, describe the erosion and sedimentation control measures which will be used. Some of these will only be temporary (i.e., for the duration of the activity or throughout construction), while others will be permanent, like seeding and sodding. The purpose of these measures is to provide stability to disturbed soils which will be especially susceptible to erosion due to high water, flooding, or rain water runoff. Stabilization should occur as soon as possible after disturbance.

You want to assure the regulators that you will minimize silt and sediment laden runoff from the site into nearby surface waters. Some minimization can be achieved by maintaining natural traps for eroded material. This can happen if clearing and grubbing, excavation and embankment operations are sequenced properly. Sedimentation barriers such as filter fabric fences and straw bales can also be used as a temporary measure.

Some temporary measures may be converted to permanent use once construction is over. For instance, sediment basins which receive rain water runoff during construction via temporary berms, dikes, and diversion ditches could also be used once operation begins, depending on how the site layout is planned.

You will also need to outline the maintenance and inspection plan to be followed during construction to ensure that erosion and sedimentation control measures are effective. This plan would include periodic inspections and subsequent reporting on temporary and permanent measures. Establish procedures for how often inspections will be conducted and the reports filed. Inspection parameters should include checking control measures such as silt fences for depth of sediment, tears, stability, etc. Once seeding has been conducted, the applicable areas should be checked periodically for bare spots. Response measures should also be provided. For example, silt will be removed from fences when the silt reaches one-third of the fence's height.

OTHER LAND IMPACTS

The land could be impacted by a variety of other construction activities. It is possible that a significant archaeological find could occur during earth moving operations. We recommend you have a written plan in place for this possibility stating that any such find would not be disturbed and that the appropriate agency (generally, the state historic preservation office) would be notified. Obtain agency concurrence with this plan.

Blasting may be required if your site will require a deep excavation for a coal car dumper or other unloading facility or if you are demolishing significant existing structures. If so, you should provide information on when, where, and how often it will occur. Blasting will have to comply with all applicable safety requirements.

Provide a plan for disposing of construction waste. This would include pieces of scrap metal, wood, other debris from construction or derived from razing existing facilities, and waste oil from equipment and vehicles used during construction. Specify whether trash and organic waste will be carried off site by a licensed disposal firm, burned, or buried. Open burning of debris may necessitate a permit from a local agency. Some metal wastes could be removed by a salvage contractor. Likewise, arrange for removal of waste oil to a permitted facility for recycling or disposal.

If you are removing soils that are unsuitable for use as foundations, or if you are removing old roadways or buildings, the material may fall under a special classification known as "Construction and Demolition Debris." This material can often be used on site as backfill or disposed to other sites (other than a sanitary landfill). On the other hand, you could discover material that has been contaminated by petroleum products, particularly if you are building an expansion on an existing power plant site. In that case, you may find yourself having to do a remediation effort. This could include off site disposal or on site treatment (e.g., "soil incineration" or bioremediation).

If you are dealing with a "brown-field" site, you should have a Phase I Environmental Site Assessment performed to verify that there are no apparent asbestos, PCB-contaminated, or lead-based paint-contaminated materials included. If such verification is not achieved, a Phase II survey is called for and specialty contractors with valid licenses will be needed to deal with these hazardous materials if they are found in sufficient quantity.

There are many other aspects of construction that could impact the land. These could include:

- effects on the stability of the soils and their bearing strength,
- sanitary waste considerations from construction force,
- on-site storage of fuels for construction vehicles,

- need for new connecting roads from local roads to site, and
- effects on the soil permeability and percolation rates.

Any other potential impacts should be investigated and included in the environmental management plan.

THE WATER

Impacts will be measured by the changes that are created to the water resources defined during the baseline assessment. At that time you ascertained the quantity and quality of the water sources within your area. These included both surface water and groundwater sources.

SURFACE WATER

Surface water impacts can occur via three means: construction activities taking place within surface water bodies, construction activities affecting discharges to surface water bodies, and through leaks or spills of compounds during construction which reach surface water bodies.

Construction activities which could take place within nearby surface water bodies include the following:

- construction of intake and/or discharge structures,
- dredge and fill operations,
- construction of barge unloading facilities,
- cofferdam placement,
- construction of piers and jetties,
- excavation of channels or canals,
- dewatering of surface water bodies, and
- installation of mitigation or enhancement measures.

These activities should be examined for possible impacts on surface water quantity (including flood protection), quality, aesthetics, navigation, and aquatic plant and animal life. Identify and describe any expected impacts in your environmental management report.

The main impact that construction will have on discharges to surface water will most likely be from stormwater runoff. The quantity of stormwater runoff expected during construction depends on many factors:

- permeability of the soils,
- amount and intensity of precipitation,
- topography of the site, and
- extent of vegetative cover.

Stormwater discharges from construction activities which disturb more than five acres must be permitted under the National Pollutant Discharge Elimination System (NPDES) program. Applicants may apply to be covered under a general permit by filing a Notice of Intent (NOI) at least two days before commencing construction at the site. Under the general permit the EPA, or the appropriate state agency in states where the authority has been delegated, requires the applicant to submit a pollution prevention plan (PPP). The PPP includes the plans for stormwater management, erosion control, and other practices designed to reduce environmental impacts from construction activities.

Guidance for preparing a PPP is included in the EPA manual entitled, "Storm Water Management for Construction Activities, Developing Pollution Prevention Plans and Best Management Practices," (EPA 832-R-92-005, September 1992). States with authority may reference their own manuals, or reference the EPA's manual. The EPA guidance document recommends the following general steps:

1. Collect site specific information.
2. Calculate runoff coefficient.
3. Select erosion control and stormwater management measures.
4. Prepare an inspection and maintenance plan.
5. Certify the PPP.
6. Submit the NOI.
7. Implement controls.

8. Inspect and monitor controls.
9. Conduct final stabilization at end of construction.

Selection, inspection, and monitoring of soil stabilization measures were discussed earlier. These measures are used in concert with stormwater management measures to control the impacts of runoff from the site during construction. Stormwater management entails shaping the topography in such a way as to facilitate runoff in a controlled manner and/or using drainage ditches to collect and route runoff to a retention or detention pond. Collection of the water in this manner allows time for sediments to settle out and prevents flooding during heavy rainfall. The stormwater is subsequently discharged to off-site surface waters, and/or allowed to percolate into the ground.

The construction stormwater management plan must include a monitoring program which meets applicable regulatory requirements. Generally, they require that certain parameters be measured within 24 hours after the occurrence of a storm of a certain magnitude (e.g., one inch). These parameters will most likely include pH, total suspended solids (TSS), and oil and grease.

The final way surface water may be impacted during construction is through leaks or spills which may migrate to a surface water body. You should prepare a Spill Prevention, Control, and Countermeasures (SPCC) Plan, such as oil-handling facilities are required to prepare under the Oil Pollution Prevention regulations (40 CFR 112), to deal with this possibility. The plan should examine what chemicals are likely to be on site and what handling and disposal methods will be utilized to control them.

Chemicals likely to be present on site during construction could include the following:

- fuels,
- fertilizers,
- detergents,
- paints,
- solvents,

- spent lubricants,
- tar,
- herbicides, and
- pesticides.

Establish a list of good housekeeping measures to be followed for their storage and handling. Ensure that proper recycling or disposal methods are required when a product is ready for disposal. Have in place a corrective action plan in the event of a spill of any regulated, toxic or hazardous material. The plan should include provisions for training on-site personnel in the proper responses to a spill.

In addition to the specific concerns listed above, there may be other control measures you need to implement to minimize water impacts at your site. For instance, if concrete trucks are to be washed out on site you should designate a lined washing area where the washwater can be collected and handled properly.

Groundwater

There are two main ways in which groundwater could be affected by construction activities: dewatering efforts and spills or leaks. Spill and leak prevention and handling was discussed above, so this section will concentrate on dewatering effects.

If dewatering is to be performed at your site during excavation activities, you will need to establish the following operating parameters:

- number of wells or well points to be used,
- pumping rate and duration,
- well diameters, and
- excavation methods.

There are groundwater flow models available which can predict the drawdown effects on nearby wetlands or wells from the dewatering operations. Provide a general site map which gives the locations of the excavation areas and the drawdown contours of the water table expected around them. You want to show what the drawdown would be with no mitigative measures, and whether that drawdown would be expected to meet applicable regulatory criteria. If not, you will have to show what mitigative measures you will take (e.g., cutoff walls, recharge systems) and how they will ensure compliance. Propose a monitoring plan to demonstrate that your predictions were accurate and that you actually are meeting any criteria that you claimed you would. If a groundwater monitoring program is already in place at the site, you may use it to monitor the groundwater during construction.

The Atmosphere

Atmospheric impacts could be caused by fugitive dust, open burning, or minor air emissions from construction equipment/operations.

Fugitive dust will probably be your chief concern during construction. The erosion and sedimentation control measures discussed earlier will assist in keeping dust particles out of the air, but there are additional measures you can take to control the problem. Dust suppression can be achieved by watering non-paved roads, parking areas, and other areas where there will be heavy vehicular traffic. Require contractors hauling sand, gravel, etc. to use dust suppression techniques including wetting of material or covers over the truck bed. If you are doing excavation or filling of earth to make level areas for your facilities, moisture control will prevent excessive dust generation. The EPA publication known as AP-42 describes assumptions that can be used to calculate estimated emissions of fugitive dust from various construction sources.

Open burning will release carbon monoxide, sulfur oxides and other chemicals into the air. If performed over short time periods, this should not constitute a severe air impact but can provoke public complaints. Ensure that burning is conducted in accordance with local regulations.

Minor emissions of volatile organic compounds (VOCs) and fuel components will occur during construction from operation and refueling of vehicles and from painting and other activities. These emissions are very low and can be minimized by using standard control measures.

The Plants

For each phase of the construction project, provide an acreage accounting and listing of what vegetation community will be impacted. If vegetation has been previously stressed (through overgrazing, fire damage, etc.), the impacts of construction should be measured against that baseline. The vegetation survey conducted during the baseline assessment will indicate if any of the vegetation to be impacted by clearing, grubbing, etc. has special significance (e.g., rare, endangered, or wetland species). On the other hand, if any of the vegetation is trash or nuisance species, take credit for eliminating it.

Your PPP should include a discussion of herbicide handling and application to prevent inadvertent destroying of desired vegetation. Herbicide use should be in accordance with manufacturer's directions and carried out by a licensed applicator. Herbicide should only be applied to targeted vegetation. Usually, herbicides are only used to maintain rights-of-way, such as under transmission lines.

The Animals

Provide a description of what impacts construction activities will have on wildlife and aquatic organisms in the site area. Include effects of loss of habitat, disruption of breeding areas, loss of vegetation which provided foodstuff, or other impacts.

Note if any rare, threatened or endangered wildlife will be impacted by the construction, or even if unusual habitat that they require will be destroyed or altered. Be sure to include impacts of noise from construction as it may disturb wildlife.

The People

Project construction can have multiple effects on the local populace, some of which are beneficial. Impacts should be measured against the socioeconomic baseline conditions established earlier.

The project will have a positive but temporary effect on employment and consequent effects on housing demand, traffic, and public services. A socioeconomics expert can predict what impacts will ripple through the local community due to the project. Construction should provide increased tax revenues for local and state governments. Often, road improvements are required to handle traffic generated by a construction work force.

Other impacts to examine include aesthetic qualities like visual impacts, odors, and noise. State in your environmental management report that these impacts will be minimized to the extent possible.

Noise impacts during construction will most likely be from heavy equipment such as backhoes or other diesel powered vehicles used in excavation operations. Typical noise levels from construction equipment are listed in Table 6–1. Noise modeling can be performed in order to predict the effect of operating this equipment on the noise receptors in the area.

Other temporary impacts due to your construction may include the following:

- overloading of water supply and sewage treatment facilities,
- crowding of local schools and hospitals, and
- increased requirements for police and firefighting personnel and equipment.

Discuss any of these potential impacts in your report.

TABLE 6–1 Example of Major Construction Equipment and Associated Noise Levels

Construction Equipment[a]	Noise Level (dBA) per Unit @ 50 ft
Caterpillar Bulldozer	74
65-Ton Crawler Crane	88
65-Ton Truck Crane	83
24-yd^3 Dirt Scraper	88
3/4 yd^3 Backhoe	85
3/4-yd^3 Front-End Loader	84
Air Compressor (250 cfm)	73
Air Compressor (750 cfm)	73
Gas-Driven Welding Units	78
Concrete Mixers	85
Concrete Pumps	82

[a] Includes only major construction noise sources greater than 70 dBA.
Source: FPL Lauderdale Site Certification Application, 1989.

Chapter 7
Effects of Operation

As mentioned in Chapter 6, the regulatory requirements for impact analysis originate in 40 CFR 1502.16, Environmental Consequences. This portion of the CEQ rules relates to Environmental Impact Statements (EISs) and calls for discussion of "any adverse environmental effects which cannot be avoided" and requires discussions of both direct and indirect effects, and their significance, and whether they are irreversible and unavoidable.

The purpose of this chapter is to advise you on how to present your assessment of the environmental impacts expected from operating the power plant to regulators. You accomplish this through field studies, computer modeling, theoretical projections or by other defensible methods. Although the regulator decides if impacts are within regulatory limits, you must show why you believe the project will meet all standards. If your investigation shows you will exceed a standard, this is the time to try to justify an exemption from the standard. This exemption is usually known as a variance.

You will probably know early on what your problem area is going to be. Don't sweep it under the rug and hope it goes away. It won't. It's better to air it out early and begin negotiating solutions than to have it come out later, when the schedule is critical. Likewise, don't ignore an aspect of the project because you think its effects will be negligible. You may have to prove it later, when you don't have time. Address the issue up front, so its importance (or lack thereof) can be determined early in the process.

Some impacts are unavoidable (e.g., the use of the site land for the life of the plant); while others can be mitigated in some way. Some impacts will outlive the project. If you are leveling a forest to develop this site, that impact will far outlast the operational stage of the project. However, so will rechanneling natural waterways to augment flow in a dying stream. Keep in mind that not all impacts are detrimental. Have some feature of your project that is a deliberate environmental benefit—wetland creation, habitat enhancement, wildlife preserve, etc. Give environmentalists something about your project which they can support. Invest effort in accentuating the positive whenever possible. Successful examples include the rehabilitation and preservation of distressed wetlands, the donation of property to wildlife corridors, and even the construction of manatee feeding/resting areas in thermal discharges.

Impact assessment is best handled by your environmental specialists with input from the engineering team. Compile their reports into an encyclopedia for future reference. Use your baseline reports as a starting point, because they define the potential impactees. Each of the impact reports should include whatever operational parameters apply (e.g., plant systems), the basis for expecting an impact (theoretical or applied methods), the operational monitoring plans for tracking impacts and any control measures you will be utilizing to limit or lessen impacts. Make sure that all of the same parameters that were assessed for the baseline in Chapter 3 are covered: the land, the water, the atmosphere, the plants, the animals, and the people.

The Land

There are three aspects of land impacts covered: land use, topography, and geology.

Land Use

The major effect of operating your plant is simply that land will be committed to it for the life of the plant, if not longer. If this use is consistent with the local zoning and land use plans, then no impacts arise from its commitment. If the land you wish to use is not zoned industrially, find out how large an effort it would be to have it rezoned and how long it will take. If at all possible, get the rezoning done before you announce your project and start meeting with agency personnel. Some zoning boards meet very infrequently (as infrequently as once a year).

If, during your baseline investigation, you discovered the presence of any "special lands" nearby, you will have to prove that operating the plant will have little or no impact on them. Such proof would come from results of noise, air quality, and/or water studies; all of which will be discussed later.

Topography

Changes in the topography of your site will start during construction when earth moving activities change the baseline contours, elevations, and relief of the land. These changes can be measured physically or by comparison with topographic maps. When new structures, man-made features and final landscaping are added for the operational phase, this is also considered a change in topography, and these changes will have an impact both on runoff patterns at the site and aesthetics. We will discuss water impacts of operation later. As far as aesthetic qualities are concerned, if you intend to add landscaped berms or to retain an effective view screen of some sort, be sure to point this out in your report. The best way to describe the changes you will be making to the viewshed is to use actual photographs of the site and to insert representations of the plant into them. Do this for each viewing direction from which the plant will be visible.

Geology

Most surface changes to the land occur during the construction phase when grading, deforestation, and earth moving take place. Major geological changes could only be precipitated by massive draining of underlying aquifers (as in Mexico City, where the structures are steadily sinking) or by formation of sinkholes (which is not unknown in some parts of the country) or by stressing with the weight from large storage or waste piles. We assume that you are not predicting any of these catastrophic occurrences and, in fact, will be able to demonstrate in your report that they will not occur.

The Water

Impacts are caused by withdrawing water, discharging wastewater into it and by runoff, leachate, or seepage into it. These impacts are site specific and depend on the systems' operational parameters and the baseline environmental conditions. Begin by examining your water withdrawal and its effect on water sources (surface and/or groundwater). Secondly, examine all of your plant systems that discharge to water. Unless you have a zero discharge system, this includes the heat dissipation system and chemical and biocide discharges. (For both withdrawals and discharges, include an assessment of aquatic lifeforms that are impacted). Lastly, look at your stormwater runoff system and other sources of influence on water such as seepage or leachate.

Water Withdrawal

One impact to local water sources may be your usage of that water. The amounts required were quantified in your plant operations report. Ascertain if these withdrawals represent a significant impact on the water sources. This includes both surface water bodies and aquifers, as applicable. Determine if groundwater drawdown will affect any of the permitted wells within a five mile radius of the site you identified during the baseline assessment. If so, determine if this violates any regulatory criteria in your area.

Surface water bodies may be affected by the presence of intake structures. Aquatic organisms identified during the baseline assessment are at danger of impingement or entrainment from the intake structures. Discuss plans for minimizing these hazards. For example, you may be able to build diversionary structures around the intake location depending on what type and sizes of fish populate the area. You can minimize the entrainment and impingement of fish by always maintaining your intake velocity below 0.5 feet per second, and by locating your intake in a place and depth where the least amounts of fish are found.

You will also have to minimize hazards to non-swimming organisms, such as:

- phytoplankton (free-floating photosynthetic plants),
- plankton (non-mobile free-floating microscopic organisms), and
- meroplankton (planktonic life stages of fish or invertebrates, usually eggs or larvae).

The only way to minimize entrainment of these organisms is to locate your intake in a place and depth where their population densities are minimized. Virtually 100 percent of entrained organisms will be killed, whether your cooling system is open or closed cycle. However, organisms will still be available as food for other species (including rare or endangered species, or commercially important or sport fish) if you have a once-through system.

WASTEWATER DISCHARGES

You will address the systems that discharge to water when you prepare a SPDES or NPDES permit application. These applications are discussed in detail in Chapter 8. Briefly, they require that you present historical sampling data for any existing outfalls and predicted data for new outfalls. These data tables are usually accompanied by a report on expected impacts of the discharges on the receiving waters and their aquatic life. The predominant plant systems which have wastewater discharges include the heat dissipation system and process and sanitary waste streams.

The heat dissipation system will both utilize water resources and affect them with its effluent. The extent of these effects depends upon the type of system used. For new systems, once through cooling systems precipitate the greatest impacts and require 316(a) and (b) demonstrations. These are named after the sections of the CWA that allow for demonstrations to be made to show that the use of once-through cooling would not have a significant adverse impact on the affected water bodies. The 316(a) demonstration deals with the discharge of heated effluent to a receiving body of water (RBW), while the 316(b) applies to the impacts of an intake on the source water body. We advise our clients early on to avoid the 316 demonstration mire and utilize either zero-discharge or closed cycle systems. Additions to an existing once-through or closed cycle system will require preparation of an environmental report to address 316 type issues.

There are two types of water standards with which you need to be familiar. They are technology based effluent limits (TBELs) and water quality based effluent limits (WQBELs). TBELs are based upon an extensive study of steam electric power plants performed for the EPA decades ago. The results of the study were published in 1974 in a document entitled "Development Document for Effluent Limitations Guidelines and New Source Performance Standards for the Steam Electric Power Generating Point Source Category" (report No. EPA-4401/1-74/029a). This report was the reference document for effluent limits published in 40 CFR 423 for steam electric power plants. TBELs are limits which have been established for commonly used systems based on the operating parameters and capabilities of those systems.

WQBELs are limits based on the quality (or the desired quality) of the receiving body of water (RBW), hence they are site specific and can be very strict. A WQBEL starts with a toxicity level for the most sensitive aquatic biota expected to be affected and back-calculates the maximum allowable discharge of the chemical that can be discharged to the RBW. Typically, the toxicity level is set to some small fraction of the lowest level of the chemical which has been demonstrated to have an adverse affect (chronic toxicity) over a long exposure time, or an even smaller fraction of the level which has a quick effect (acute toxicity). The flow in the RBW is usually set to a very small flow, such as the seven-consecutive-day, ten-year low flow for a river. Since WQBELs are calculated without regard to practical limitations of technology, it is common for them to be even lower than the detection limit for a chemical. Typically, a WQBEL will be based on chronic toxicity for the RBW outside some designated mixing zone, and based on acute toxicity within that mixing zone.

We had a client who intended to install a new discharge to an already dead river (acid mine drainage) using the TBELs in their existing permit as operating limits. However, the state biologist discovered a small habitat of aquatic organisms in the receiving stream and instigated very strict WQBELs instead. This required our client to add more expensive wastewater treatment equipment than was present in the original design. They spent a great deal of effort (synonym: money) pursuing a variance to avoid this and in the end were unsuccessful.

The RBW may be the same as the source water body but is subject to a different set of impacts from discharges. The first will be thermal effects. The addition of heat load will increase the temperature of the system. Computer modeling is an effective way to predict the spatial and depth extent of the expected thermal plume. Present the results of such modeling (see Figure 7–1 for an example) in the report along with details on its derivation, calibration and operation. Based on rules originally proposed by the EPA, regulators generally look for plume extent from the maximum temperature rise all the way down to the one degree Fahrenheit isotherm. Depending on your location, they may want to see plumes for spring and fall, but will always want to see them for winter and summer.

For an existing plant, use historical records of temperature monitoring to demonstrate effects (or lack thereof) from the existing system. If your existing system includes a cooling pond, you may be able to offset the expected temperature increase by cleaning the pond of aquatic vegetation, thus enhancing its ability to transfer heat. Hopefully, you were clever enough to design your cooling pond so it would not be considered waters

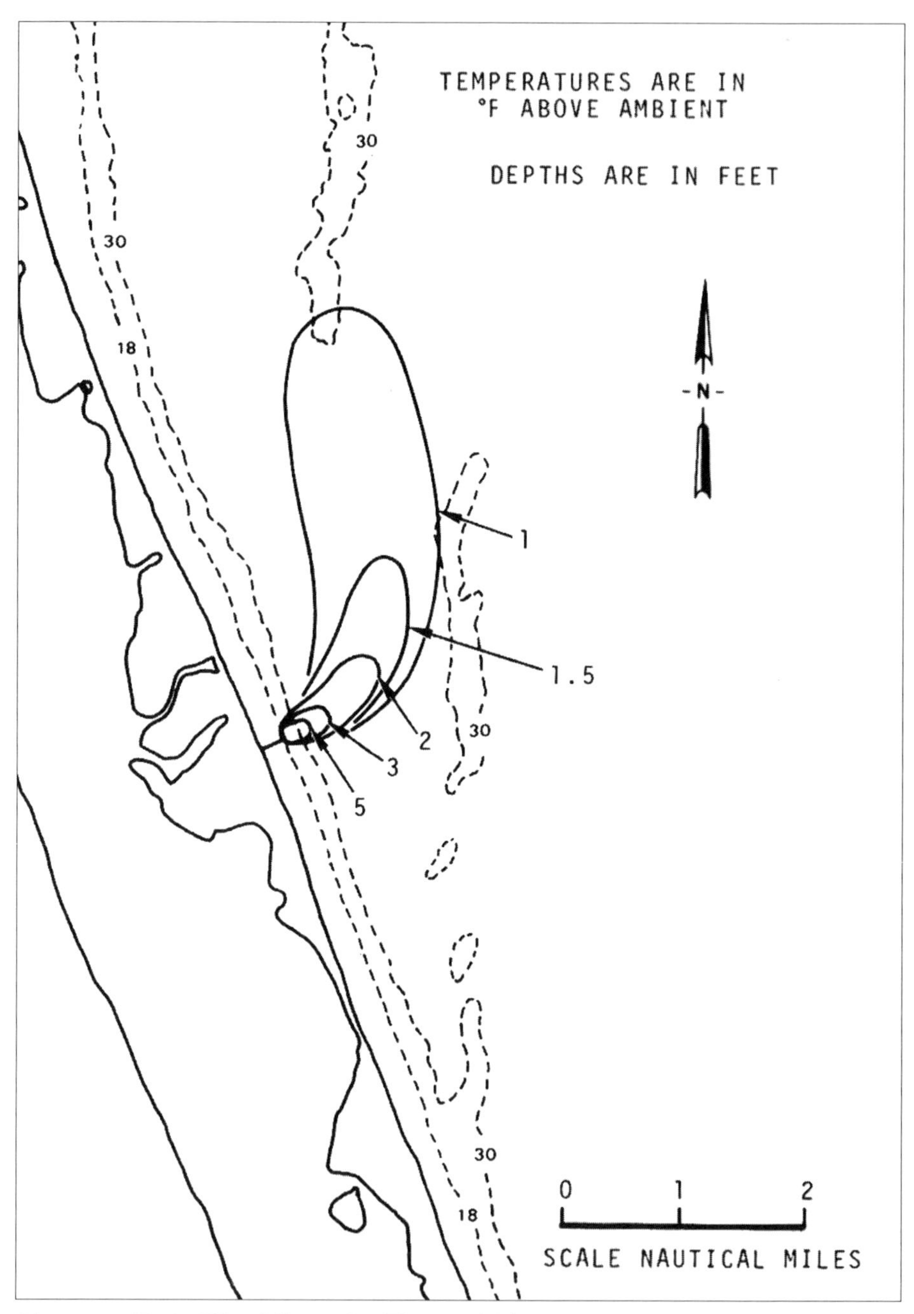

Figure 7-1 Plan View of a Thermal Plume

of the U.S. or of the state you are in. The general rule is that your pond is a cooling lake and waters of the state if it impounds the main stem of a river, and normally releases water downstream that has arrived from upstream.

If your pond was constructed by raising a dam all the way around a flat area, or by damming a minor (intermittent) tributary, it is not waters of the state but considered a wastewater treatment facility, much like a cooling tower. We know of one case in which a cooling pond that was not waters of the state or the United States was actually made waters of the county, but such an incident is rare. If your pond is waters of the state, you will need a mixing zone in it for your own discharge, and a dredge and fill permit anytime you need to do maintenance dredging in it. If it is not waters of the state, your thermal plume and maintenance activities are your own business.

Beware of going too far to try and get public opinion behind your project. At least two utilities have promoted their cooling ponds as multipurpose reservoirs, and even gone so far as to allow agencies to stock them with fish. They were quite surprised (and had to spend a lot of money) when they were given temperature limits within their own cooling ponds.

Aquatic life may be impacted by the presence and cessation of a thermal plume and by changes wrought by recirculation of its habitat waters. Based on the species of organisms and fish present, estimate the effects of the thermal plume on these creatures. Don't forget to examine impacts from cessation of the thermal plume due to shutdowns (this is called "cold shock") on creatures which have grown accustomed to the warmer temperatures—manatees, for instance. Finally, your biologist should determine the overall impact of withdrawing from and resupplying the water source which serves as the home for aquatic organisms. Discuss any chemical (e.g., dissolved oxygen) or nutrient level changes that may occur from recirculating the waters. Use a well-documented thermal model to predict these impacts by superimposing chemical concentration onto it.

Present a well-thought-out monitoring plan for measuring physical and chemical characteristics of the effluent and receiving bodies of water during operation of your project.

Chemical and biocide discharges from your process and sanitary wastewaters will also be identified and quantified during the NPDES/SPDES application process. Your biologist should determine and discuss the effects of these discharges on aquatic life and vegetation, such as sea grass, in the receiving water body, based on predicted levels. This includes the acute and chronic toxicity, long-term bioaccumulation in sediment, effects on the food chain, and comparisons with the LC-50 values for the organisms affected. One methodology for presenting this information is to use contour lines of equal concentration compared to a contour line of the applicable water quality standard concentrations drawn on a map of the receiving body of water.

STORMWATER RUNOFF, LEACHATE, AND SEEPAGE

The drainage basin areas which are feeding each water body were ascertained during the baseline assessment phase. Plant operation will affect the drainage areas and patterns at the site in various ways. Also, removal of drainage areas reduces the flow in receiving streams. Address the possible impacts of changing drainage patterns and flows, and reference all applicable regulatory criteria and how you intend to meet them. Look at the impacts leachate or seepage could have on water resources. There may be some significant interaction between the ground and surface systems via hydraulic connections. This is particularly important for plants utilizing cooling ponds because seepage may affect both groundwater and surface water levels and water quality in the area. Your hydrologist should ascertain what impacts will be expected.

THE ATMOSPHERE

The atmosphere may be affected in several ways by operation of your plant. There may be impacts to meteorological phenomena, and there will most likely be impacts on the ambient air quality from plant emissions.

Meteorological Phenomena

Determine if operation of your heat dissipation components will affect the meteorology at the site. It can do so through generation of fog from a cooling pond, cooling tower drift, or visible vapor. If such fog is likely, determine its impact on local transportation routes (both on the ground and in the air). If domestic wastewater is to be utilized as cooling water, analyze the potential biological and health impacts of the cooling tower drift. Determine if cooling tower drift could contribute to icing of nearby roads during the winter. If you are in a coastal area and have a salt water cooling tower system, you can probably show that salt deposition from cooling tower drift is less than ambient natural salt deposition. Locate your cooling towers so drift will not deposit on your switchyard or high voltage transmission lines, and so the warm and moist air leaving one tower is not sucked into another one, thus impairing its performance. Figure 7–2 is an example of how to present predictions of salt deposition.

Air Quality

You will address the impacts of operating your plant on ambient air quality in the Prevention of Significant Deterioration (PSD) permit application. The PSD application will be discussed in detail in Chapter 8. During the application process you will have to demonstrate that operating the plant will not cause significant impacts on air quality because you cannot violate PSD increments or air quality standards. You will not only have to show that you are meeting air quality standards set to protect human health, but also that you will not have any effect on Class I areas.

Your report should contain all of the information necessary for the PSD application and include your planned monitoring program to monitor air quality during operation.

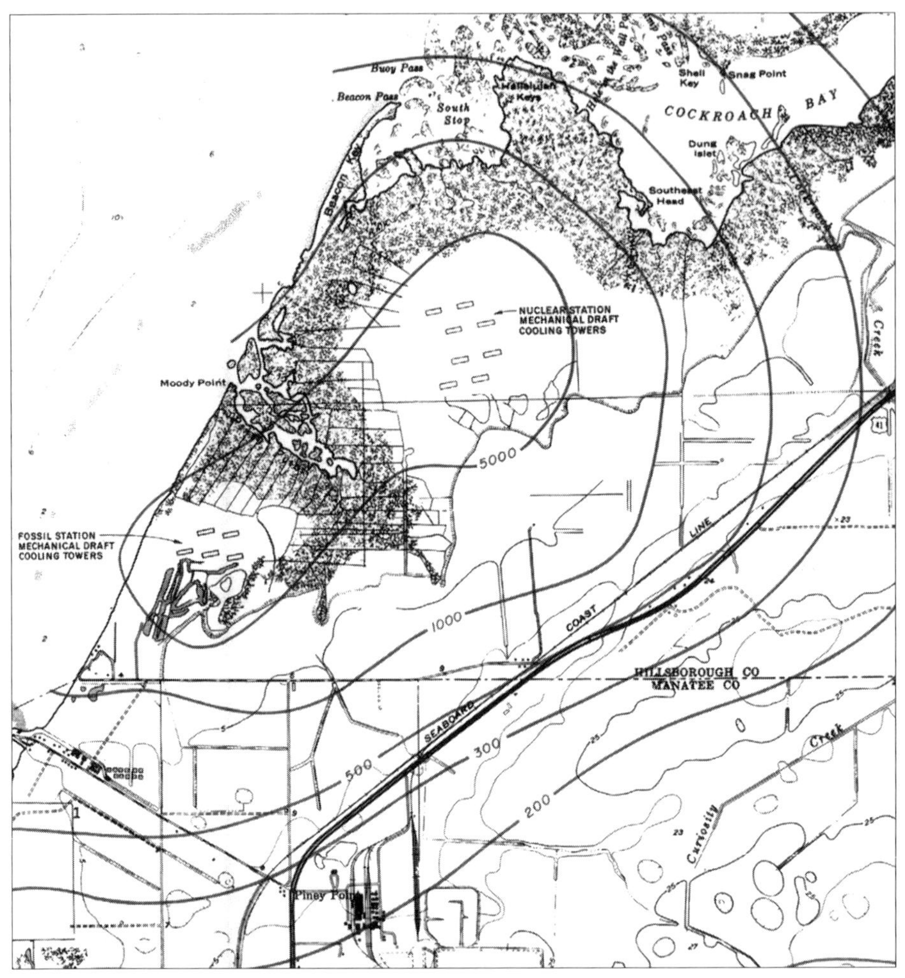

FIGURE 7-2 Predicted Salt Deposition in Pounds per Acre per Year

THE PLANTS

During the baseline assessment, your biologist identified the major vegetation types at the site and vicinity and the presence, abundance, and condition of the vegetation. Any impacts of operating the project on these habitats should be presented in a report. Possible impacts include:

- air toxic effects on vegetation,
- effects of cooling tower drift or fugitive dust emissions,
- traffic that could damage off-site habitats, and
- effects of drawdown on surface or groundwater that could dry out wetlands.

Of chief importance will be the presence of any species which are considered threatened, endangered, or under some other type of regulatory protection.

THE ANIMALS

The terrestrial ecology survey performed during the baseline assessment characterized the wildlife populations of the existing site and vicinity in enough detail that the regulators can be sure the project will not harm or disturb species of importance. Any impacts of operating the project on animals should be presented in a report by your biologist. Possible impacts could include the effects of cooling tower drift and noise, and increased road kills due to increased traffic.

THE PEOPLE

Any impacts of plant operation on the labor force, housing market, public services, transportation system, and the aesthetic quality of the existing site (noise and visual qualities) should be presented in your report. Such impacts could include:

- impacts of truck or rail deliveries to the site,
- impacts of increased traffic around the site, and
- changes in roads near site (addition of turning lanes).

This is one area where benefits can be emphasized. If your socio-economic expert did an adequate job during the baseline assessment of characterizing the local community, it should be relatively simple to gage the potential impacts of the plant. These include the following possible positive elements:

- generation of additional tax revenues for local government,
- construction or improvement of roads,
- creation of new jobs, both temporary and permanent,
- boost to the local economy from wage-earning workers, and
- creation of parks or other enhancements available for public use.

It is difficult to quantify some of these positive impacts but you should make an effort to determine how many people would benefit from them.

On the flip side, you must examine possible adverse socioeconomic impacts of the project. This could include more traffic congestion, housing shortages, overloading of public utilities (e.g., schools, hospitals, police, fire), and aesthetic qualities like noise and visual impact.

If there is a plan to dispose of waste materials off-site, calculate the weight of material expected to be disposed and determine how this will affect the life of the off-site landfill. You will also have to show the amount of the incremental increase in truck or rail traffic that will be required to move this material, as well as any deliveries of raw materials such as fuel oil, coal, or limestone and what impact this increase in traffic will have on the roads and the people.

CHAPTER 8
PERMITS

There will probably be a number of specific federal, state, regional, and/or local permits which you will have to obtain, particularly for a new plant. The regulatory basis for most permits comes from federal laws passed by Congress and then turned into regulations by the EPA. Most of these programs are set up to be eventually transferred to the states. While the states and localities can have stricter standards than those promulgated by the EPA, they cannot have more lenient ones. We will discuss two major permitting areas, Clean Water Act (CWA) permits and Clean Air Act (CAA) permits.

CLEAN WATER ACT

The regulatory basis stems from the Water Quality Act of 1965, the Federal Water Pollution Control Act as passed in 1972, and as amended by the Clean Water Act in 1977, and by other differently named Acts after that, but most commonly referred to as the CWA. Congress required the states to examine all stream segments within their jurisdiction and define water quality classifications for them (CWA Section 303). Classifications are based on the possible uses of the water, whether for drinking water, for protecting the aquatic life, or for agricultural, recreational, or some other specific use. Each classification was assigned ambient standards which would ensure the integrity of the intended use (CWA, Section 302). These ambient standards are used to set water quality based effluent limits (WQBELs), which are usually enforced after allowing for a mixing zone within the Receiving Body of Water (RBW).

The CWA is composed of six sections or Titles, as follows:

- I Research and Related Programs,
- II Grants for Construction of Treatment Works,
- III Standards and Enforcement,
- IV Permits and Licenses,
- V General Provisions, and
- VI State Water Pollution Control Revolving Funds.

Under the CWA, the major efforts are usually either a National Pollutant Discharge Elimination System (NPDES) permit, a 316 Demonstration, and/or a Dredge and Fill (10/404) Permit.

NATIONAL POLLUTANT DISCHARGE ELIMINATION SYSTEM (NPDES)

Regulatory Basis

The CWA established a permit program for point source discharges of industrial effluent into national waters and established effluent standards so the ambient water classifications could be maintained or restored (CWA, Section 402). This is the NPDES permitting program. The Act required states to submit implementation plans outlining intended programs for ensuring compliance with the provisions. Once approved, the state could take over administration and enforcement of the NPDES program. Most states now have authority to do so and their programs are known by the moniker of SPDES, or State Pollutant Discharge Elimination System.

In those states that have not been delegated NPDES authority, the EPA still maintains the program. The following discussion references the EPA program, but each state with NPDES authority has an equivalent program with equivalent forms and processes.

EPA was required to establish standards of performance for both new and existing industrial sources by source (CWA, Sections 302 and 306). 40 CFR 423 contains the rules for steam electric power plants. It identifies typical power plant wastewater streams and sets limits on what EPA perceived to be the worst pollutants included in those streams. A permit may not be issued unless the facility utilizes effluent treatment technology that achieves those limits without benefit of any mixing at the end of the pipe. End-of-pipe limits based on the technology available to remove the pollutant in question are TBELs.

EPA also has published a list of toxic water pollutants for which they have or are developing compound-specific effluent limitations. These are typically based on acute toxicity, which is a level that can cause immediate damage to organisms. The purpose of these limits is to avoid acutely toxic levels of pollutants within mixing zones. The WQBELs are intended more to avoid chronic toxicity levels in the rest of the RBW, outside the mixing zone.

The TBELs were established to ensure that facilities were providing a level of treatment consistent with the state-of-the-art of wastewater treatment. In other words, a discharger to a large water body cannot shirk on treatment requirements because there is a large volume of water available for mixing (dilution).

In 1990, EPA promulgated regulations for NPDES permitting of non-point sources of stormwater discharges from industrial sites (40 CFR 122.26). There are three permitting mechanisms under the regulations. One mechanism is the group application whereby plants with similar discharges can submit data from only a subset of the group's facilities and not all of them. Other applicants can submit an individual application, or file a Notice of Intent (NOI) to be covered under a General Permit. Coverage under a General Permit requires the applicant to prepare a Storm Water Pollution Prevention Plan (SWPPP), and to update it periodically.

A power plant will typically try to incorporate stormwater outfalls from a separate stormwater NPDES permit into their individual wastewater NPDES permit the next time that the permit comes up for renewal.

Application Process

The NPDES application is due 180 days prior to startup of a new facility (i.e., discharge) or, for an existing plant, 180 days before the existing permit expires.

The NPDES application for a power plant usually requires submittal of two forms. The first form is called Form 1 and asks for general information including the owner, location, contact person, Standard Industrial Code (SIC) 4911, and a site location topographic map. Form 1 actually is part of the EPA Consolidated Permits program, and is required for anyone who needs an NPDES permit, a RCRA (Resource Conservation and Recovery Act) permit, a UIC (Underground Injection Control, under the Safe Drinking Water Act) permit, or a PSD (Prevention of Significant Deterioration, under the Clean Air Act) permit. Typically, a power plant will need the NPDES and PSD permits, but not the RCRA or UIC permit.

Form 1 also asks a series of questions to clarify which supplementary forms are required. Supplementary forms are for specific facilities/discharges as follows:

- Form 2A POTWs;
- Form 2B Facilities which include animal feeding or aquatic production;
- Form 2C Existing facilities;
- Form 2D Proposed facilities (other than those covered by Form 2A or 2B);
- Form 2E Non process wastewaters;
- Form 2F Stormwater discharges;
- Form 3 Facilities which treat, store, or dispose of hazardous wastes;
- Form 4 Facilities where certain injections of fluids occur in support of oil or natural gas recovery; and
- Form 5 Any of 28 industrial categories (including steam electric plants greater than about 30 MW) which could potentially emit >100 tons per year of certain air pollutants in or near an attainment area.

Generally, a power plant only submits the Forms 1 and 2F, either 2C or 2D, and 5. The Form 5 will be discussed under the CAA. The 2C or 2D forms are to discharge wastewater and require the same type of information. Information for a proposed facility is based on best engineering estimates, since the discharge does not yet exist. Information for an existing facility must be based on measured values. Information requested in the Form 2C and 2D is as follows:

- latitude, longitude, and receiving water name of each outfall;
- water flow diagram/balance;
- description of all operations and average flow rate of each stream contributing to each outfall;
- description of any treatment undergone by each stream described above;
- for intermittent streams, the frequency, duration, flow rate, and volume;
- information about effluent guideline limitations;
- any improvements required by regulatory compliance letters, court orders, or other applicable enforcement documentation;
- list of expected pollutants and data for them;
- biological toxicity testing data (if any);
- laboratory information; and
- certification as to accuracy and truthfulness of information presented.

Data cannot be older than three years, but only one sample value is required. If the permitting agency feels it is not representative, it may ask for additional sampling. If you are renewing an existing permit, this is not a problem because your old permit is automatically extended as soon as you file the new application. If, however, your application is for a new source or discharge, and you have a definite schedule to meet, you should submit data from at least three sampling episodes, each at least a month apart. Provide data for all of your outfalls, and also for the ambient intake water. If a pollutant is present in the ambient intake water at concentrations greater than the water quality criteria allow, most states have a mechanism for setting alternative (higher) criteria.

The Form 2F requires the following information:

- latitude, longitude, and receiving water name of each outfall;
- any improvements required by regulatory compliance letters, court orders, or any other applicable enforcement documentation;
- site drainage map;
- description and acreage (pervious and impervious) of drainage areas;
- description of material treated, stored, or disposed in contact with stormwater runoff;
- control measures in use at each outfall;
- description of any leaks or spills of toxic or hazardous pollutants in past three years; and
- sampling data and corresponding storm event.

The Permit

In issuing permit limits, the regulator relies on the measured or predicted data the applicant submits in the application. The permit will most definitely have limits for what are called conventional pollutants: pH, dissolved oxygen, thermal effects, fecal coliform, oil and grease, and other aesthetic properties. Other limitations will be based on the pollutants demonstrated by the sampling data or expected by engineering estimates. The trend now is toward WQBELs that are more stringent than TBELs. This trend is caused by the ever-lowering sampling detection limits, and the associated ever-lowering levels at which toxic effects on aquatic organisms are being documented.

The issued permit will be effective for five years.

316 Demonstration

Regulatory Basis

The regulatory basis for the 316 demonstration lies in the CWA, which includes under Title III a Section 316, entitled Thermal Discharges. Section 316(a) allows effluent limitations for the thermal component of discharges to be made less stringent than the effluent limitation specified by EPA for power plants. That limit is essentially that no heat can be discharged by once-through cooling systems. Section 316(b) addresses intake effects on a water body. Section (a) requires that a plant "demonstrate" that permitting of the thermal discharge will "assure the protection and propagation of a balanced, indigenous population of shellfish, fish, and wildlife in and on" the RBW. Section (b) requires a similar demonstration be performed to show the "location, design, construction, and capacity of cooling water intake structures reflect the best technology available for minimizing adverse environmental impact."

The guidelines developed by the EPA related to thermal discharges are in 40 CFR 122. Detailed regulations for type (a) demonstrations are listed in 40 CFR 125.70 through 125.73 (Subpart H). The corresponding regulations for 316 (b) demonstrations should have been codified in Subpart I of 40 CFR 125, but have not been.

EPA has published two manuals in conjunction with the NRC to postulate the actual data requirements that should be met to make a complete 316 demonstration. They were published by the EPA Office of Water Enforcement, Permits Division, Individual Permits Branch, on May 1, 1977. They are entitled "Interagency 316(a) Technical Guidance Manual and Guide for Thermal Effects Sections of Nuclear Facilities Environmental Impact Statements" and "Guidance for Evaluating the Adverse Impact of Cooling Water Intake Structures on the Aquatic Environment : Section 316(b) P.L. 92-500."

Application Process

A 316(b) demonstration includes a biological study of the intake water body and a prediction of the entrainment and impingement effects of the plant's circulating water system on the aquatic life. The demonstration document should include the following:

- site location and meteorology;
- hydrology (flow and temperature records, bathymetry, and sediments) of intake water body;
- ecological assessment of the intake water body (aquatic life survey);
- any existing environmental stresses;
- circulating water system components, design, and operation;
- intake structures; and
- potential for impact to the aquatic life from impingement and entrainment.

A 316(a) demonstration includes information on the design of the discharge facility and the configuration of the receiving body of water (RBW). The demonstration should include the following components:

- site location and meteorology,
- plant description,
- hydrology of RBW,
- ecological assessment of the RBW (aquatic life survey),
- circulating water system,
- discharge outfall configuration and operation,
- expected thermal plume,
- potential for cold shock, and
- potential for thermal effects on aquatic life.

The Permit

There is no actual permit issued for a successful 316 demonstration. Typically, a sentence is added to the NPDES Permit to the effect that a successful type (a) and/or (b) demonstration was submitted and is held as part of the permitting agency's files. The wise utility will keep copies of both demonstrations and any correspondence relating to them, in case the permitting agency should lose track of those files.

Dredge and Fill Permit

The dredge and fill permit can be the most confusing of the CWA permits. Strictly speaking, the dredge and fill permit is not just a CWA permit, but also falls under the Rivers and Harbors Act of 1899 (RHA). Section 9 of the RHA requires that any bridges, causeways, dams, or dikes constructed in or over navigable water of the United States have to be approved by Congress, and that the plans for such have to be submitted to and approved by the "Chief of Engineers and by the Secretary of the Army." Section 10 of the RHA is even more strenuous, and deals with requirements for the same review by the Chief Engineer and the Secretary of the Army for "The creation of any obstruction ... to the navigable capacity of any of the waters of the United States."

The Army Corps of Engineers (COE) has its own Code of Federal Regulations, entitled Navigation and Navigable Waters, and labeled 33 CFR. The following parts of 33 CFR are all relevant to dredge and fill permits:

- Part 320 General Regulatory Policies,
- Part 321 Permits for Dams and Dikes in Navigable Waters of the United States
- Part 322 Permits for Structures or Work in or Affecting Navigable Waters of the United States,
- Part 323 Permits for Discharges of Dredged or Fill Material into Waters of the United States,
- Part 324 Permits for Ocean Dumping of Dredged Material,
- Part 325 Processing of Department of the Army Permits,
- Part 326 Enforcement,
- Part 327 Public Hearings,

- Part 328 Definition of Waters of the United States,
- Part 329 Definition of Navigable Waters of the United States, and
- Part 330 Nationwide Permits.

Meanwhile, under section 404 of the CWA, EPA also promulgated 40 CFR Subchapter H, entitled "Ocean Dumping." Some of these regulations are relevant to Ocean Dumping (Parts 220 through 229), including a section entitled "Corps of Engineers Dredged Material Permits" (40 CFR 225). 40 CFR Parts 230 through 233, although still part of Subchapter H, actually deal with the waters of the United States that are not the oceans, including disposal of dredged materials (Part 230), procedures (Part 231), 404 Program definitions (Part 232), and 404 state programs (Part 233).

Because they are based on section 404 of the CWA, the EPA rules only apply to the discharge of dredged materials. Using the RHA, the COE has a much broader-based scope of regulation. It includes not only the discharge of dredged materials, but also includes any activities that may affect the navigable waters of the United States (33 CFR 322). As a result of this distinction, EPA and COE reached an agreement (actually a Memorandum of Understanding or MOU) by which the COE normally administers the dredge and fill permit, but utilizes the guidance of EPA in making decisions. Because the authority is from both Section 10 of RHA and Section 404 of CWA, these permits are sometimes called 10/404 permits.

In many states, there are also state and/or local dredge and fill permit requirements, which are separate and different from the federal requirements. However, the agencies often cooperate. For example, in Florida, there is a joint form entitled "Joint Application for Works in the Waters of Florida." This form requires the following information:

1. Applicant's name and address.
2. Name, address, zip code, telephone number, and title of applicant's authorized agent.
3. Name of waterway at work site.

4. Location of work.
5. Names, addresses, and zip codes of adjacent property owners who also adjoin the water.
6. Proposed use.
7. Desired permit duration.
8. General permit or exemption requested.
9. Total extent of work.
10. Description of work.
11. Turbidity, erosion, and sedimentation controls proposed.
12. Dates of start and finish of activity.
13. Previous applications for this project.
14. Certification that applicant controls the property.

Other states and other COE districts may have variations on the form, but will generally ask for the same information. Typically, applicants make use of multiple attachment documents to explain and illustrate the proposed activities. The COE requires that all applications include no paper larger than 8.5 by 11 inches in size. This requirement forces large projects to use multiple sheets to present maps and aerial photographs.

Clean Air Act

Regulatory Basis

The regulatory basis for the air permitting process lies in the Clean Air Act (CAA) which was originally passed in 1970. The Act has been amended over the years, most recently by the 1990 Clean Air Act Amendments (CAAA). As noted in Chapter 1, the CAA set national ambient air quality standards (NAAQS) for major air pollutants and required states to develop state implementation plans (SIPs) to implement and maintain these standards. A state can have stricter rules than the federal program, but not more lenient ones. Upon approval of their SIP by EPA, the state has authority to administer the program. At this time, the regulator of the clean air program in most cases is the state. EPA does keep oversight authority, and has actually rescinded SIP approval in some cases.

EPA published regulations to codify the CAA, as amended, in 40 CFR Parts 50 to 81. Most states have incorporated these rules, by reference, into their own regulations. The regulations include official definitions, standards, monitoring methods, and compliance and administration information. They introduced a host of officially defined acronyms which rival the stars in number. There are other officially defined terms which we will place in quotes so that you can look up the definition. The federal regulations which outline permitting and standards most applicable to power plants are as follows:

- 40 CFR 50 Primary and Secondary NAAQS,
- 40 CFR 60 New Source Performance Standards, and
- 40 CFR 61 National Emission Standards for Hazardous Air Pollutants (NESHAPS).

The CAAA of 1990 totals almost 800 pages. They are composed of 11 sections or Titles, as follows:

I Attainment & Maintenance of National Ambient Air Quality Standards (NAAQS),
II Mobile Sources,
III Hazardous Air Pollutants (HAPs, also called Air Toxics),
IV Acid Deposition Control (acid rain),
V Permits,
VI Stratospheric Ozone Protection,
VII Enforcement,
VIII Miscellaneous Provisions,
IX Clean Air Research,
X Disadvantaged Business Concerns, and
XI Job Displacement Provisions.

Basically, the CAAA introduced a uniform national permitting process with enforcement authority, market-based approaches to meeting limits, and incentives for using cleaner fuels, reducing energy waste, and implementing conservation programs. They specifically addressed the growing problems of urban air pollution (smog, CO, and PM_{10}), acid rain, and global warming due to ozone depletion. As of publication of this book, EPA is in the process of codifying the provisions of the CAAA into regulations. The program is in a state of flux, whereby some aspects have been challenged in court and others are under a phasing in schedule over the next decade. Most deadlines hinge around the year 2000, but these could change.

Air permitting is the most complex of the permitting efforts. If you have to do any air permitting, we recommend you hire a competent, experienced consultant who can keep up with the constantly changing regulations. However, for your information, here are the major points of the air program and where things lie at the present.

Congress designated certain areas of the country—mostly large national parks and wilderness areas—as pristine Class I areas, which are to receive the utmost protection from air quality deterioration. The rest of the country falls under Class II, which means that it is subject to less stringent requirements than Class I areas. EPA, with input from the states, examined air quality throughout the country for the levels of "criteria air pollutants," such as ozone, CO, and PM_{10}. The levels were compared to the NAAQS, which are primary standards designed to protect human health. There are also secondary NAAQS which are designed to protect the environment (e.g., buildings, plant life).

Areas of the country which did not meet the NAAQS for a pollutant were designated as nonattainment areas for that pollutant. Those areas which were meeting the NAAQS were designated as attainment areas. Thus, the country became a giant jigsaw puzzle where a piece could be attainment for one pollutant but not for others. The nonattainment areas were further classified based on the severity of the pollution problem. The sub-classifications for ozone nonattainment are: marginal, moderate, serious, severe, or extreme. CO and PM_{10} nonattainment areas are sub-classified as either moderate or serious.

An industry which wants to locate a major new source or modify an existing major source within a nonattainment area goes through a preconstruction process called "New Source Review." The industry is held to stricter emission standards than a similar industry in an attainment area. These stricter standards are called the Lowest Achievable Emission Rates (LAERs). LAERs are the stricter of the most stringent emission rates contained in any state's SIP or those actually achieved by a same or similar source in use somewhere. They are listed in 40 CFR 61. Note: it is similar to the Reasonably Available Control Technology (RACT) which is usually applied to existing sources in nonattainment areas.

To locate a major new source or modify an existing major source within an attainment area, an industry also goes through a "New Source Review" called the Prevention of Significant Deterioration (PSD) application process. This is a decision-making process in which the regulator first determines whether the addition or modification will cause significant deterioration of the ambient air quality. PSD permits require that the Best Available Control Technology (BACT) be installed for each pollutant subject to regulation. This is a top-down approach whereby the applicant first examines the best control technology on the market for that source to see if it is a reasonable choice for their plant based on economic, environmental, and energy considerations. If it is not the best choice, the applicants can work their way down to the technology which is the best choice for the plant. Thus, BACT differs on a case-by-case basis and is determined with the regulator's concurrence.

A major new source to be built in an attainment area must meet New Source Performance Standards (NSPS), which are stricter than standards for an existing source in the same area. NSPS have been established for specific sources or industries and are listed in 40 CFR 60 Subparts D through PPP. Most power plant sources fall under the following subparts:

- D Fossil-Fuel Fired Steam Generators,
- Da Electric Utility Steam Generating Units,
- Db Industrial-Commercial-Institutional Steam Generating Units, and
- GG Stationary Gas Turbines.

The BACT chosen for a plant must ensure that the emissions will meet the NSPS. The state can issue one permit which incorporates PSD, the application of BACT, and compliance with the NSPS.

There are official definitions of "new source," "major stationary source," and "major modification" in the CAAA. The classifications depend on your plant's location (attainment or nonattainment area), the type of source and how many tons per year of pollutants the source has the "potential to emit" (PTE). This latter term is defined as the maximum amount which the source could emit (after application of control technology), not the actual expected amount. A "major stationary source" is one which exceeds certain thresholds of pollutants as shown in Table 8–1.

One of the most progressive ideas presented by the CAAA involves emission banking and trading. If you are planning to build a new plant in a nonattainment area, you can do so if you offset the new emissions by reducing emissions from an existing source somewhere within the area. This could be accomplished by retiring the existing source or by applying better control equipment on it. The net result is an overall reduction in emissions. A facility which substantially reduces the emissions from an existing source can "bank" credits for later use or sell them to another facility. This market-based approach allows some flexibility in how the overall emissions limits are met, particularly for utilities which own several power plants within an area. A plant where control technology costs are cheaper can supply an offset for another plant which is having difficulty meeting the limits or for a new source.

Title III covers HAPs or air toxics, chemicals which are known or suspected to cause cancer, mutations, developmental or neurological disorders, or other acute health problems. Title III included a list of 189 HAPs (although this list has been subsequently reduced), and EPA classified certain industries as "major sources" for HAPs. These industries will have to use maximum available control technology (MACT) to meet the National Emissions Standards for Hazardous Air Pollutants (NESHAPS). A facility is classified as a "major source" if it has the

TABLE 8–1 Threshold Levels for Major Stationary Sources

Pollutant	NonAttainment Status	Threshold (tpy)
Criteria/NSPS	Attainment	100
VOC/NO_x	Marginal	100
	Moderate	100
	Serious	50
	Severe	25
	Extreme	10
VOC	All other areas in	50
	ozone transport region	
CO	Serious	50
PM	Serious	70

[SOURCE: 40 CFR 70.2]

"potential to emit" 10 tons per year of an individual HAP or 25 tons per year total of HAPs. The emission offset program also applies to HAPs.

Title IV contains provisions to reduce acid rain (mist, snow, fog, dust, etc.). This Title requires reductions in SO_x and NO_x by power plants covered by the program. Allowances or SO_2 tickets will be issued to these facilities. Each allowance is worth one ton of SO_2, and each covered plant will be issued allowances which total less tonnage than they are currently generating. Thereafter, the plant cannot emit more SO_2 than its allowances cover. Plants which achieve significant reductions can sell their extra allowances to other plants or to a broker. Allowance trading is also allowed. All covered power plants will have to use continuous emission monitoring system (CEMS) to monitor SO_x and NO_x. Under Title IV, plants can achieve bonuses for using clean coal technology, encouraging consumers to conserve electricity, and by other means.

The new national air permitting process, similar to that of the NPDES program for water emissions, is contained in Title V. It requires industries to do an inventory of existing or potential emissions to determine if a Title V permit is required. The permit is good for five years. The PSD permit can serve as the Title V permit for new sources or those undergoing major modifications. This permit is the equivalent of the EPA Form 5 discussed under Clean Water Act above, or the state equivalent.

Application Process

In general, here is the information required for the Title V application:

- process description and control equipment performance;
- emissions inventory (including comprehensive quality data);
- any existing permit conditions;
- applicable federal, state, and local regulations;
- operating scenarios and worst case conditions;
- expected process changes during term of permit;
- potential for emissions trading;
- monitoring plan; and
- compliance plan and schedule.

To learn more about what is required in a Title V application, read 40 CFR 70.5(c). A PSD application will most likely be used as the basis for Title V permit applications. Although most states have their own PSD application form, the application usually includes a report with the following contents:

- introduction,
- project description,
- air quality review requirements and applicability,
- Best Available Control Technology,
- ambient air quality monitoring data analysis,
- air quality modelling approach,

- air quality impact analysis results,
- additional impacts analysis, and
- references.

CHAPTER 9
INTERNATIONAL PERMITTING

Permitting a power plant in a different country can be either more or less difficult than permitting one in this country. On the one hand, very few countries in the world have more stringent environmental regulations and requirements than the United States. On the other hand, the culture, political system, and social customs are likely to be completely different. To ensure that you know the environmental requirements of the host country, as well as acceptable cultural, political, and social methods of operation, we strongly recommend you hire a local consultant.

Having said that, there are two main ways you might become involved in the permitting of a power plant in another country: as a builder or as a buyer. You may or may not also be the operator.

The main types of power plants with which you can become involved internationally are nuclear, hydroelectric, and fossil fuel-fired. If the plant is nuclear, you will not have to worry about air emissions, but will have to worry about radiation safety issues (which are outside the scope of this book). You will have to worry about water and wastewater issues, as for all steam electric plants, but these issues are similar whether your fuel is nuclear or fossil.

If the plant is hydroelectric, you will not have to worry about air emissions or wastewater issues, but you will have to deal with the impacts of construction (destroying habitat by impoundment) and operation (changing river flow rates, temperatures, and durations; and blocking fish migration). These will likely be multiple use projects, including such uses as agricultural irrigation for example, and the environmental analysis will have to be done upfront as a fatal flaw analysis. The institutional framework is similar to that described below for fossil fuel-fired plants.

If you are permitting a power plant internationally, the odds are that it will be a fossil fuel-fired plant. The remainder of this chapter will be based on that assumption.

NEW POWER PLANTS

The majority of international power plant permitting is associated with the construction of new generating units in Asian and Latin American countries trying to develop an infrastructure similar to that in the United States. Have your consultant identify all of the applicable host country regulations, if there are any, and put you in touch with all of the responsible government entities. It is important that you reach an agreement with them as to what environmental (and other) requirements you will have to meet. You should also make an effort to discuss your project with representatives of any environmental non-governmental organizations (NGOs) likely to become involved. Again, your local consultant will be able to identify these for you.

You will probably obtain financing from one or more lending institutions which will probably insist you meet their own environmental requirements, just to safeguard their loan. The World Bank is a multi-lateral lending institution, with its own Environment Department. The department has developed environmental guidelines (Operational Directive 4.01) on how to do an Environmental Assessment (EA) to determine whether a project is acceptable.

The World Bank is one of the largest lending institutions in the world, and many host countries and other financial institutions use its guidelines when approving power projects. The World Bank published a draft Industrial Pollution Prevention and Abatement Handbook (the Handbook) in collaboration with the United Nations Industrial Development Organization and the United Nations Environment Programme in 1995. This handbook contains industry-specific detailed requirements for several industries, including one entitled "Fossil-Fuel Based Thermal Power

Plants." Anyone trying to permit construction (or major modification) of a power plant in a developing country can expect to have to make a detailed showing of following the Handbook.

The Handbook has been issued with the intention of complying with the principle of sustainable development, in which industrial development and environmental protection are kept compatible. In compiling the Handbook, the World Bank staff appear to have examined the environmental regulations of three separate institutions, and selected the ones that they felt were most appropriate. The three institutions are:

1. World Health Organization,
2. U.S. EPA, and
3. European Union.

Air Emissions

The Handbook identifies three critical air pollutants for which minimum standards are listed. These are particulate matter (PM), sulfur oxides (SO_x), and nitrogen oxides (NO_x). It requires that the baseline values for PM_{10} (particulate matter of diameter less than 10 micrometer), SO_x, and NO_x be established first without the project and that pre-existing sources be identified. It requires that "an appropriate dispersion model" (e.g., an EPA-approved model) be used to predict concentrations resulting from the project. It also sets emission limits for those pollutants which are not to be exceeded and target levels of treatment which are to be achieved. The emission limits, minimum standards, and target levels are all to be met unless you can demonstrate that others are more appropriate (either higher or lower). The emission limits, standards, and target levels are shown in Table 9–1.

The Handbook also requires you to provide continuous emissions monitoring for particulates, SO_x, NO_x, and heavy metals "as appropriate," and fuel ash and sulfur content.

TABLE 9–1 Air Emission Minimum Standards and Target Levels

Pollutant	Emission Limit	Standard	Target Level
PM	50 mg/Nm^3	80 ug/m^3 (annual avg.)	99% removal[a]
PM_{10}	Not Applicable	50 ug/m^3 (annual avg.)	98% removal[a]
NO_x (coal)	650 mg/Nm^3 (230 ug/J)	200 ug/m^3 (400)[c]	40% removal[b]
NO_x (oil)	360 mg/Nm^3 (100 ug/J)	200 ug/m^3 (400)[c]	40% removal[b]
NO_x (natural gas)	240 mg/Nm^3 (65 ug/J)	200 ug/m^3 (400)[c]	40% removal[b]
SO_x	2000 mg/Nm^3	80 ug/m^3 (250)[d]	Not Applicable
SO_x	<0.2 tonnes/day/MW for first 1000 MW, plus 0.1 tonnes/day/MW above 1000 MW	80 ug/m^3 (250)[d]	Not Applicable

[a] 95% of the time that the unit is running

[b] relative to no controls, 95% of the time that the unit operates

[c] not to exceed 7 hourly exceedances in 1 year (for any hour)

[d] annual average (not to exceed 7 24-hour average incidents in a year)

LIQUID EMISSIONS

The handbook also sets liquid effluent limits, although it does not set water quality standards in the receiving body of water (RBW) except for temperature. The effluent limits not to be exceeded are as follows:

pH	Between 6.0 and 9.0 standard units
TSS	50 mg/L
Oil & grease	10 mg/L
Total residual chlorine (TRC)	0.2 mg/L
Total chromium	0.5 mg/L
Hexavalent chromium	0.1 mg/L
Copper	0.5 mg/L

Iron	1.0 mg/L
Nickel	0.5 mg/L
Zinc	0.5 mg/L
Temperature Rise	3 degrees C at the end of a mixing zone (100 meters from the discharge unless otherwise defined)

These effluent limits are less restrictive than the U.S. EPA limits in 40 CFR 423, and much less stringent than any WQBELs in the United States. The Handbook requires continuous monitoring of pH and temperature, and monthly sampling of TSS, TRC, heavy metals, and other pollutants for which you have a treatment system.

Solid Wastes

The Handbook also specifies requirements for solid waste disposal for ash and scrubber sludge. If the waste is dewatered ash or chemically stabilized FGD sludge, disposal can be to one of the following:

- ordinary landfill, if soils are impermeable and groundwater is deep;
- lined cells, if seepage is a concern; or
- mines, if risk of contamination of surface or groundwater is "appropriately managed."

If the sludge is unstabilized, the handbook calls for leachate collection and control. By implication, disposal of ash that is not dewatered is not permissible.

Key Issues

The Handbook identifies what it calls certain "key issues." These include using the cleanest fuel available, balancing environmental and economic benefits, careful ash disposal, and comprehensive monitoring and reporting. It recommends the following:

- recirculating cooling systems,
- low NO_x burners,
- dry FGD systems, and
- high levels of particulate removal.

In reality, the Handbook's preference for clean fuels rather than pollution cleanup technology is probably based upon economic reasons rather than environmental ones, as it is not compatible with the stated goal of sustainable development. All in all, however, the Handbook appears to require reasonable balance between environmental and economic concerns. You will almost certainly have to deal with it if you are to permit a new plant or major renovation (e.g., a repowering) of an existing plant.

The choice of fuel will often be made for you by the host country's utility or government, based on either economic or national security reasons. In such a case, you will then have to design the plant and its environmental control systems around the chosen fuel.

EXISTING POWER PLANTS

If you are going to permit an existing plant in another country, the chances are that you bought it. You may even have purchased an overseas utility. In either case, you probably have experience operating and permitting power plants in the United States. The types of permitting you must do include renewals of permits that have expired or addressing compliance achievement plans if you have significant violations. In other words, you will be dealing with the same type of permitting of existing plants that you deal with domestically.

If you are experienced in domestic permitting, you undoubtedly have evolved an environmental management system (EMS) that allows you to proactively manage all of the permits for your plant(s). If you don't have a formal system, you probably have frequent violations and management by crisis. In that case, you do have an EMS, but it is an inefficient and informal one. As environmental regulations become more restrictive, you will have to have a formal EMS.

The International Organization of Standardization (ISO) is an organization dedicated to achieving consistency within the global economy. This is inevitable as the world grows smaller and smaller. The ISO has been trying for hundreds of years to convert the entire world to the metric system. This group has anticipated the globalization of the worldwide power industry that is presently under way. The current trend toward deregulation of the electric utility industry in the United States is part of that globalization.

ISO has promulgated an environmental management standard called ISO 14001. This is essentially a blueprint to use in setting up an EMS if you don't have a formal one, or it is a means of obtaining independent third part certification of your EMS if you do have one. Other sections in the 14000 series will deal with the actual activities that are required, but 14001 sets the management framework.

ISO 14001 has the following six parts:

Part 4.0—General

This part directs you to establish and maintain a formal EMS structure within your company organization chart and in the job descriptions of the appropriate people within your organization.

Part 4.1—Environmental Policy

This part directs you to have a written formal policy statement, signed by management and publicly displayed. The policy includes a commitment to employee training and continual improvement in environmental compliance and pollution prevention.

Part 4.2—Planning

This part addresses environmental aspects of your facility (e.g., plant mass balance, wastewater disposal, air emissions, solid/hazardous waste disposal practices, community right-to-know, site assessments), legal aspects (confidentiality of self-audit results), objectives and targets, and a management program to assign a team, allocate manpower and establish a schedule and set of deliverables.

Part 4.3—Implementation and Operation

This part addresses the structure and responsibility (funding and its documentation), initial and periodic training for personnel, feedback to management, rewards or consequence for performance or non-performance, means of communication (electronic or paper) to satisfy regulatory requirements, documentation and document control (findable permits, plans, etc.), operational control (taking data and reporting results as required), and emergency response training and procedures.

Part 4.4—Checking and Monitoring

This part deals with monitoring, non-conformances and how to correct them, record requirements and retention times, and auditing of your EMS.

Part 4.5—Management Review

This part requires management review from the top of the organization, including its schedule, agenda, and review of action items (from last review).

American companies which implemented the ISO 9000 series on quality management found that setting and utilizing a formal system results in improved efficiency and reduced costs to the companies. Simply put, it makes common sense to implement such a system to keep your company out of trouble. Using the ISO 14000 series makes especially good sense if you are operating internationally because it is accepted internationally. Governments will take its inclusion as a sign that your are trying to do the right thing, and will probably fine you less for violations for trying.

Here is the bottom line. If you intend to operate a power plant anywhere in the world, you need to install an EMS in order to stay ahead and stay proactive on your environmental compliance issues. Once the EMS has been set up, it will enable you to stay in compliance and renew your permits as required with as little expense as possible.

Index

A

B

C

D

E

F

G

H

I

L

M

N

O

P

R

S

T

U

V

W

Y

Z